Morphological Intelligence

Keyan Ghazi-Zahedi

Morphological Intelligence

Measuring the Body's Contribution
to Intelligence

 Springer

Keyan Ghazi-Zahedi
Max Planck Institute for Mathematics
Leipzig, Sachsen, Germany

ISBN 978-3-030-20623-9   ISBN 978-3-030-20621-5   (eBook)
https://doi.org/10.1007/978-3-030-20621-5

This Springer imprint is published by the registered company Springer Nature Switzerland AG
The registered company address is: Gewerbestrasse 11, 6330 Cham, Switzerland

To Rieke, Maren, and Marion

Acknowledgements

This book would not have been possible without the support of many people. First of all, I sincerely thank Nihat Ay for his support and critical remarks. This book would not have been possible without him. I thank Martin Bogdan for his support of my Habilitation at the University of Leipzig. I also thank the Max Planck Institute for Mathematics and the Sciences for providing me with the resources that were essential to completing this book, as well as the Santa Fe Institute for hosting me during the initial phase of writing it. I want to thank my collaborators, without whom the results presented in this work would not be possible: Guido Montúfar, Oliver Brock, Daniel Häufle, Raphael Deimel, Syn Schmitt, and Johannes Rauh. Last but not least, I want to thank Antje Vandenberg for her constant support over the last years.

Funding

This work was partly funded by the German Priority Program DFG-SPP 1527 "Autonomous Learning." This publication was also made possible through the support of a grant from the John Templeton Foundation. The opinions expressed in this publication are those of the author(s) and do not necessarily reflect the views of the John Templeton Foundation.

Contents

Chapter 1
From Morphological Computation to Morphological Intelligence

You can't do much thinking with your bare brain.

Daniel Dennett

When I feel thirsty, I grasp a glass of water and move it to my mouth without thinking about the exact size and material properties of the glass. When I run through the woods, I can relax my mind, because I don't have to concentrate on the ground that I am running on. The question that motivates this book is: How is this possible? Why don't I have to carefully estimate the shape, size, and material properties of the glass to avoid breaking it or letting it slip out of my hand? Why don't I have to constantly monitor and precisely estimate the unevenness of the ground that I am running on to prevent myself from stumbling? The answer to these questions is that the physical properties of my body reduce the amount of computation that my brain has to do. In the case of grasping the glass, the softness and friction of my skin permit for some variation of the pressure that I use to hold the glass. The elasticity of the muscle-tendon system in my legs compensate for the unevenness of the ground, and hence, allow me to run with a considerably large variation of the feet placements. Given the time that the neural signals require to travel from the feet to the brain and back, running on uneven terrain would be impossible if it had to be carefully controlled by the brain. To summarise both cases, the body allows us to think less in situations that otherwise would require extensive and expensive calculations. Stated more simply, the body reduces the cognitive load or computational demand for the brain. Understanding and quantifying how the physical properties of the body reduce the cognitive load for the brain is the goal of this book.

The importance of the body with respect to intelligence and behaviour was first explicitly demonstrated by Brooks [1] with his robot Ghengis. Brooks was able to show that a complex morphology can be controlled in real-time with limited onboard computation, which at its time (the mid-80s), was groundbreaking. In subsequent publications, Brooks [2, 3] wrote that the "*source of intelligence is not limited to just*

© Springer Nature Switzerland AG 2019
K. Ghazi-Zahedi, *Morphological Intelligence*,
https://doi.org/10.1007/978-3-030-20621-5_1

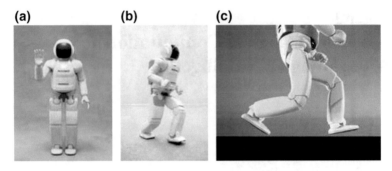

Fig. 1.1 *Asimo*. Images were taken from [6]. From left to right: **a** Honda's Asimo. **b** Asimo running. **c** Running for humans is commonly defined as a locomotion behaviour which has phases during which both feet leave the ground. This plot shows that Asimo can run

the computational engine." Although this seems to be a simple statement, it is the main difference between embodied artificial intelligence and artificial intelligence, or GOFAI (good old-fashioned AI, [4]). GOFAI, which currently is the dominant approach in e.g. robotics, focusses mainly on the computational engine (brain) and considers the body (if it is considered at all) as a source of noise that needs to be controlled and compensated for examples see [5]. Brooks argues that "*it [intelligence] also comes from the situation in the world, the signal transformations within the sensors, and the physical couplings of the robot with the world.*" This coupling between the brain and the environment through the embodiment is today known as the sensorimotor loop. Furthermore, Brooks states that "*the intelligence of a system emerges from the system's interactions with the world and from sometimes indirect interactions between its components.*" To summarise these statements in more recent terms, Brooks is saying that intelligence is a phenomenon that arises from the interaction of the brain, body, and environment and that it cannot be assigned to any one part of it. It can only be understood and modelled if all parts of the sensorimotor loop are taken into account at all times. This approach is today known as embodied (artificial) intelligence.

The difference between embodied artificial intelligence and GOFAI or GOFR (good old-fashioned robotics) is best illustrated along with the two examples of Asimo [6] and the Passive Dynamic Walker [7, 8].

Honda's Asimo (see Fig. 1.1 and [9]) is the most-advanced humanoid robot [6]. It is fully controlled in the sense that the movement of every joint is closely monitored and controlled at every point in time. This also includes carefully controlling feet placement and ground contact forces. The method is known as zero-moment point algorithm (ZMP, [10]). ZMP allows impressive running with speeds up to 9km/h, where running means that there are phases during which both feet have no contact with the ground (see Fig. 1.1c).

The Passive Dynamic Walker [7, 8, 12] is a purely mechanical system (see Fig. 1.2) that resembles human legs. The legs are carefully designed with respect to their length and weight proportions (inspired by human legs) and its feet are carefully shaped for

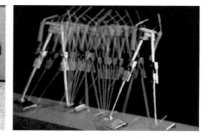

Fig. 1.2 *Passive Dynamic Walker*. Left-hand side: Image of the Passive Dynamic Walker. Right-hand side: Display of the resulting walking behaviour. All images are taken from [11]. The images show the morphology of the system (left-hand side) and the walking pattern of the Passive Dynamic Walker (right-hand side) when it is placed on a slope

walking. If it is placed on a slope with a specific decline, it is able to walk without any control at all. The walking behaviour is the result of the physical interactions of the system (length and weight distribution, feed shape, etc.) and the environment (slope, friction, gravity, etc.).

Contrary to Asimo, the Passive Dynamic Walker has a very small niche in which it shows the naturally appealing walking behaviour. If any parameter is changed, e.g. the shape of the feet, it will not walk but simply fall over. Nonetheless, it shows that walking can occur, at least partially, uncontrolled. In particular, the swing phase, i.e., the phase in which the hind leg swings forth to initiate a new step, can occur completely uncontrolled in human walking.

The question now is, what happened to the computation that is required for Asimo in the case of the Passive Dynamic Walker? An intuitively plausible answer is to say that the control has now being taken over by the physical interactions of the body and environment. This has lead to the notion of *Morphological Computation*, which is best described in the following quote:

> *Morphological Computation refers to the computation which is conducted by the body, that otherwise would have to be performed by the brain* [13].

This was initially described as *morphology and control trade-off* [14] and is arguable one of the most central concepts in the field of embodied artificial intelligence today.

Although there is a lot of agreement that the Passive Dynamic Walker is a good example for morphological computation (as discussed e.g. in [13, 15–23]), there are also publications which argue against it. Nowakowski [24], Müller and Hoffmann [25, 26] argue that the passive walker is not computing. Füchslin et al. [27] argue more specific that the Passive Dynamic Walker is missing the programming aspect of morphological computation. We will address all of these points throughout this chapter and revisit the Passive Dynamic Walker and the concept of Morphological Computation. Before we enter this discussion, it is helpful to understand the different ways in which a morphology and its interaction with the environment contribute to intelligence and behaviour. This will be done in the following paragraphs.

We are not the first to point out various forms of morphological contributions to cognition. A distinction between sensor morphology taking over computation from the brain, and the body's shape and materials which interact with the environment was made in [18, 28]. Müller and Hoffmann [25] suggested to distinguish between morphological computation, morphological control, sensor pre-processing, and physical processes which do not contribute to a behaviour. The introduction of the latter is particularly interesting, because it explicitly highlights that there are physical processes which enable behaviour but do not reduce the cognitive load for the brain. In this book, we build upon the previously made distinctions and add two forms of contributions of the morphology to intelligence which were not discussed in previous publications.

Hence, the next sections will discuss *Morphological Computation, Morphological Control, Pre-processing (sensors), Post-processing (actuators), Brain layout*, and finally *Behaviour-enabling physical processes*. These are not strictly disjunct categories. For example, pre-processing in sensors can also be understood as morphological computation in the way it is described in the next section (see Sect. 1.1.1). The goal of this book is to develop a unifying theory for all types of morphological contributions to a behaviour. It is therefore not problematic if specific examples fit into more than one category.

1.1 How Morphology Lifts the Computational Burden for the Brain

In this section, we describe different ways in which a morphology can reduce the amount of computation that the brain has to perform to produce a specific behaviour. The section starts with morphological computation as it was historically introduced and covers the transition to the version in which it is largely understood today. We also investigate other ways in which morphology reduces the computational cost for the brain or controller. These include sensor pre-processing, actuator post-processing, and brain layout. Finally, we contrast this with physical processes that enable behaviour, but which do not contribute to a reduction of computational effort.

1.1.1 Morphological Computation

The term *Morphological Computation* was first introduced by Pfeifer and Bongard [13] and Paul [29]. Paul [29] explains the original concept along the XOR robot, which is a robot that has two inputs, *A* and *B*. The robot's behaviour is best described as moving = A XOR B. What is interesting about the XOR robot, is the fact that the XOR cannot be found anywhere in the robot's controller. The two inputs are connected to the two motors by logical AND and OR circuits.

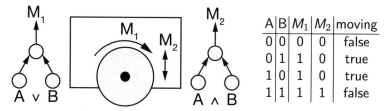

A	B	M_1	M_2	moving
0	0	0	0	false
0	1	1	0	true
1	0	1	0	true
1	1	1	1	false

Fig. 1.3 *XOR Robot.* Left-hand side: Schematics of the XOR Robot. Image was redrawn from [29]. The robot's observable behaviour is a given by the XOR of its two inputs (see the table of the right-hand side), although the XOR is not implemented as part of the controller. It is a result of the robot's morphology and it's interaction with the environment. If $M_1 = A$ OR B is true, the wheel is rotating. For the robot to move, the wheel must also touch the ground, for which $M_2 = A$ AND B must be false

To understand how the XOR is implemented, the morphology, which is described in detail next, has to be taken into account (see Fig. 1.3). The two binary inputs A and B control two motors M_1 and M_2, which both actuate a single wheel (see Fig. 1.3). The first motor M_1 controls the rotation of the wheel, while the second motor M_2 controls the wheel's position. The input to each motor is a boolean function of the two inputs. In particular $M_1 = A$ OR B (rotation of the wheel) and $M_2 = A$ AND B (position of the wheel). Hence, the wheel turns if either $A = 1$ or $B = 1$, and it touches the ground if either $A = 0$ or $B = 0$ ($M_2 = 1$ means that the wheel is lifted from the ground). The result is that the robot only moves if A XOR $B = 1$ (see Fig. 1.3, right-hand side), although the XOR function cannot be found in the control program. One can argue that the XOR is computed by the morphology, which lead to the term *Morphological Computation*. The XOR robot is a well-designed thought experiment that nicely visualises the concept of morphological computation.

Puppy is a four-legged robot that shows how body-environment interactions can perform computations in a real robotic system (see Fig. 1.4 and [30–33]). The authors state that their goal is to *"explore design principles of the whole body dynamics for the purpose of sensing."* Puppy is an under-actuated four-legged walking robot with springs in its legs. It is interesting in this context because its body dynamics (walking pattern) results from the interplay of a simple open-loop controller, morphological properties (e.g. springs), and environmental properties (e.g. friction). The authors argue, that if *"a system exploits morphological properties (e.g. shape, stiffness, friction, weight distribution), it is possible to simplify the control architecture and to achieve energy-efficient behavior."*

In the next step, the authors investigated Puppy's passive body dynamics with respect to the generation of information that could be used as sensory input. This was first discussed by Lungarella et al. [34] and Pfeifer et al. [35], who showed that sensor-motor coupling can lead to an information self-structuring. In this context, Puppy's body dynamics are understood and analysed in the same way. Comprehensive analysis of Puppy's body dynamics revealed that the sensor information can be used to determine environmental parameters, such as the texture of the ground. This is a real-world version of the XOR robot. Instead of calculating the XOR function

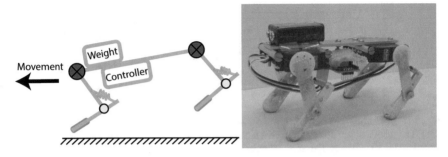

Fig. 1.4 *Puppy*. Left-hand side: Schematics of Puppy, redrawn from [36]. Puppy has four active joints (two are shown in purple), four passive joints (two are shown in yellow) and four springs (two are shown in orange). The robot is controlled in open-loop, and its behaviour is determined by the texture of the ground and the feet materials (shown in brown). Right-hand side: Puppy (image was taken from [36])

through the morphology, Puppy calculates information about the environment through its body-dynamics.

More recently, there is the notion of morphological computation in which the morphology is considered as a physical reservoir computer. Reservoir computing [37–41] is a branch of machine learning that allows to learn complex functions by only learning an output layer of a recurrent neural network. If the structure of the recurrent neural network is chosen correctly, meaning that it consists of several loops with different length and appropriately chosen weights, the dynamics of the neurons' outputs will converge to attractors with different periods. In short, a recurrent neural network will have a rich dynamical repertoire. In reservoir computing, these dynamics are harnessed by only training a read-out layer of neurons. In the context of morphological computation, the rich reservoir of dynamics is not provided by a recurrent neural network, but by the body and its interactions with the environment. There are numerous examples which demonstrate the computational power of soft robotics, in particular, the computational power of an artificial octopus arm placed in water (see Sect. 1.1.1 and [19, 42, 43]). For example Nakajima et al. [44] show that the dynamics of the soft robotic arm can be used to reproduce a non-linear dynamics system, in particular, a non-linear auto-regressive moving average (NARMA) system.

Another interesting example of morphological computation form biology is fish. Zambrano et al. [15] show that a "*dead fish is propelled upstream when its flexible body resonates with oncoming vortices formed in the wake of a bluff cylinder, despite being well outside the suction region of the cylinder. Within this passive propulsion mode, the body of the fish extracts sufficient energy from the oncoming vortices to develop thrust to overcome its own drag.*" In short, this quote states that fish exploit body-environment interactions for efficient swimming. This is also observed by others, e.g. [45], who show that "*salmon and trout have been found to entrain behind obstacles to optimize their net energy expense.*" cited from [46].

Liao et al. [47, 48] showed that live trout synchronise in both, frequency and phase, with the incoming wake for energy benefit. This shows that the morphology of fish not only reduces the energetic cost for swimming [45, 47, 48], but that it is also partly responsible for swimming motion, as shown for the dead fish [15]. The effect was investigated and reproduced by [49, 50], who showed that foil can produce thrust in wavy streams. They also showed that it is easier for flexible fish to extract energy from the environment than it is the tin foil. This is very similar to the Passive Dynamic Walker, which extracts energy from the environment (gravity) to walk down the slope. In a similar way, fish exploit body-environment interactions for swimming.

The next concept, *Morphological Control* is used to describe systems, which utilise the output of *Morphological Computation* for control.

1.1.2 Morphological Control

Morphological control was already discussed by Paul [29], where it was called *explicit* morphological computation. The difference between implicit and explicit morphological computation is that implicit morphological computation can only be observed by studying the behaviour while the output of explicit morphological computation is fed back into the controller or brain [29]. Implicit morphological applies to all systems described in the previous section. As systems with explicit morphological computation use the computations conducted by the morphology as inputs to the controller. This is now known as *Morphological Control* [25–27].

One example for such a system is the spine-driven four-legged robot built by Zhao et al. [51]. The robot has an elastic spine that is compressed and stretched by a central pattern generator. The spine-dynamics are also influenced by the body-environment interactions, which in this case, means the interactions of the feet with the ground. Hence, the current state of the spine, e.g. the amount of compression, is the result of the robot's controller and the result of the recent behavioural history of the robot, i.e., how the body interacted with the environment.

The spine is equipped with a series of sensors that measure its compression at different points. The outputs of the sensors are fed into a single-layer neural network which generates control signals for the motors. The equivalence to reservoir computing is now obvious. The outputs of neurons in a recurrent neural network are replaced by sensors reporting the state of the body. In both cases, a single-layer is then trained to produce a desired output from a reservoir of dynamics. This last step is the important distinction between Morphological Computation and Morphological Control.

1.1.3   Pre-processing (Sensors)

There are impressive examples that demonstrate how sensor placement and sensor morphology reduce the amount of computation that the brain or controller has to conduct. Sensing is described by [52] as the change of material properties *"in response to one or more external stimuli, including acoustic, electromagnetic, optical, thermal, and mechanical."* The authors state further that

> the physical properties of the material itself affect not just sensing and actuation, but also computation. Indeed, material dynamics allow one to shift classes of computation such as feedback control (e.g., by exploiting thermal or chemical deformation to regulate a process), rectification [e.g., to compensate for motion parallax in an insect's eye [53]), or transformation of a signal into the frequency domain (e.g., in the cochlea in the inner ear), by simply tuning the geometry and material properties of a structure.

To summarise the quote, sensors transform external stimuli into neural signals that are then used for computation or control. This is the reason why sensing and sensor placement are considered as pre-processing. The second interesting part of the quote describes how changes in the physical properties affect sensing, and hence, how pre-processing is influenced by the morphology. The example given in the quote is the ear's cochlea. The frequencies to which the cochlea is sensitive change with its geometry.

Another example of sensor-preprocessing are whiskers, e.g. in rats. Several studies show that rats actively scan their environment with their whiskers to discriminate between object shapes and object textures [54]. The fine-grained mechanical stick-slip motions of the whiskers on the surface of an object [55] enable easy discrimination of object textures, e.g. by training a simple neural network [56]. The geometry of the whiskers are important [57] as well as spatial frequencies [58]. To summarise, the interaction of the physical properties and the spacial arrangement of the whiskers result in neural signals that allow for the easy discrimination of object shapes and object textures. This has been reproduced in robotic experiments [59, 60]. Similar effects were exploited in the context of texture identification for a soft finger with randomly placed sensors [61].

1.1.4   Post-processing (Actuators)

Analogous to the way in which sensors enable a reduction of computation for the brain by pre-processing external stimuli, actuators (e.g. muscles, wings, etc.) can interact with the environment in such a way that it reduces the computational complexity for control. We call this post-processing because these contributions of the morphology to intelligence occur after the motor signal was generated by the brain or controller. The following quote by Wootton [62] is an example which shows that there is an uncontrolled part of the interaction between an insect's wing and the environment that contributes to the wing's function:

> [...] active muscular forces cannot entirely control the wing's shape in flight. They can only interact dynamically with the aerodynamic and inertial forces that the wings experience and with the wing's own elasticity; the instantaneous results of these interactions are essentially determined by the architecture of the wing itself [...]

A similar principle to the one described by Wootton is essential also in the context of micro air vehicles [63]. In their work, the authors describe the design, fabrication, and analysis of a 3cm wingspan flying robot which is loosely based upon the morphology of insects of the order Diptera. The interesting aspect of this artificial insect is that it relies on passive joints in its wings to fly. The wing's second and third degree of freedom which control the rotation of the wing, are passive in their design. The passive rotation of the wings is essential for the lift of the robot [63].

Another example in this section are tensegrity robots, which are often cited and investigated in the context of morphological computation. The term tensegrity was initially coined by Buckminster Fuller [64, 65] as a linguistic blend of the two words tension and integrity. Tensegrity structures consist of two types of elements, those in pure compression (mostly rigid elements such as rods) and those in pure tension (mostly elastic elements, such as ropes). Elements of the same group are not allowed to be directly connected to each other. This means that a rigid element is only directly connected to tension elements and vice versa.

The smallest possible tensegrity structure is shown in Fig. 1.5 (left-hand side). The other two structures shown in Fig. 1.5 illustrate the general idea of tensegrity robots. For this purpose, imagine that the length of one or more rods in each of the three structures could be manipulated. Let us assume that the same number of rods are manipulated in their lengths with the same frequency and amplitude in each of the three structures shown in Fig. 1.5. It is obvious that the behaviours, which result from the length modulations will be different for the three structures, as they e.g. have different resonance frequencies. Furthermore, the position(s) and number of the actuated rod(s) influence the observed behaviour for each of the structures individually. This has spiked the interest of numerous researchers, investigating the interaction of the morphology with very simple actuation, mainly oscillation of one imbalance (see e.g. [29, 66–70]). Oscillating an imbalance means that a single motor with an imbalance is attached to a rod and set into rotation. As a result, the entire structure is set into vibration. The observed behaviour will vary depending on the frequency of the rotation and the location of the imbalance in the structure.

Tensegrity robots are not just academic but interesting in terms of applications because they are robust in several ways. First, the inherent elasticity results in robustness with respect to impact. This is one reason, why NASA is investigating these structures for extra-terrestrial exploration [71–73]. Second, tensegrity robots can be easily constructed redundantly, which means that failures only affect local parts of the robot [74].

In a study related to tensegrity robots, Reis et al. [17] systematically investigated the influence of morphological modifications on the behaviour of simple robots. In their study, they bent steel to an inverted u-shape. The result is a minimalistic robot with elastic properties. Each robot has a motor which rotates an imbalance at the centre of the hip. Variation of leg and hip lengths were evaluated and mathematically

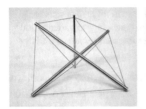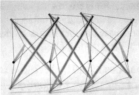

Fig. 1.5 *Tensegrity structures*. Images by Kenneth Snelson (http://kennethsnelson.net) From left to right: X-Piece, Wing II, X-Column Variation. X-Piece is the smallest possible tensegrity structure, consisting of three rigid and 6 elastic elements

modelled. The authors discussed in detail how variations of the morphology resulted in different walking patterns. Although structurally less complex compared to tensegrity robots, they are very similar. In both cases, we find a simple actuation that leads to a variety of behaviours as a result of body-environment interactions.

Muscles are another good example of post-processing of motor commands. There are several models of muscles found in literature (see e.g., [75, 76]). Haeufle et al. [77, 78] have investigated the required control effort for muscle models with increasing complexity and found that more complex muscle models require less control effort [79]. In a follow-up study, we quantified morphological computation of three hopping models (see also Sect. 5.2 [80]). These studies demonstrate the contribution of muscle dynamics to a behaviour, and hence, are examples of post-processing in muscles.

There are numerous other studies of post-processing in natural systems, such as flexible spines that are essential for the safe landing of falling cats [81], patterns of movement employed in locomotion [82–86], energy efficiency in running [87, 88], self stabilising mechanisms [89–91], and most efficient mode of locomotion [86, 92–98], which can all be considered as a form of morphological post-processing.

This section closes with a study by Dermitzakis et al. [99], who showed that friction in the muscle-tendon system of human fingers is beneficial as it contributes significantly to the total output force (approx. 9–12%). A similar effect is known from bats, who use friction in their muscle-tendon system to lock their fingers without the application of muscular force [100–102], and hence, without the need of actively controlling the grasp.

1.1.5 Brain Layout

The main point of this section is that the structure on which the computation is carried out has an influence on the efficiency of the computation. The term computation will be discussed in more detail below (see Sect. 1.2.3). For the purpose of this section, it is sufficient to think about computation as a process that transforms some sort of input into some sort of output on a technical device (e.g. pocket calculator) or a biological structure, e.g. a brain.

In the context of this section, we are interested in the question, if e.g. the neural pathway from sensor to motor neurons in the human brain is of such nature that it reduces the computational cost for certain tasks.

Tabareau et al. [103] have investigated this question with respect to the superior colliculus (SC), which is a layered structure in the brain that controls gaze orientation. They were able to prove mathematically, that the logarithmic mapping of the neurons in the SC of monkeys contributes to the transformation of the SC spacial code (activation patterns due to sensor stimulus) to the saccade burst generators (SBG) Cartesian temporal code. This is also known as the spatio-temporal transformation (STT). In other words, by mapping the input logarithmically onto the neurons of the next layer, computation has been outsourced to the specific layout and connectivity of the neural network structure. This has been shown to reduce the computational effort for the brain by Tabareau et al. [103].

A related example can be found in the context of computer science. Today, every central processing unit (CPU) is equipped with multiple caches, e.g., L1 and L2 cache, which reduce energetic and time cost for the CPU. The reason is called *The Principle of Locality* [104]. This term captures the fact, that memory references for instructions and data tend to cluster over time in two ways. First, *Temporal Locality*, which means that if the CPU accesses a memory location M it is very likely that it will access the same memory in the next computation cycle. Second, *Spacial Locality*, if the CPU accesses a memory location M, is very likely to access a neighbouring memory location in the next time step. Benchmarks have shown, that 90% of the execution time is spent in about 10–15% of the code [105], which is a strong argument for spatial locality. Hence, localised memories (caches) were introduced early in CPU chip design. The cache is a smaller, faster memory, which is located very close to the CPU, i.e., where the data is needed. It stores data and instruction that are frequently used, and hence, improves the performances and energy efficiency of modern CPUs.

1.1.6 Behaviour-Enabling Physical Processes

The previous sections discussed how physical properties in the morphology and body-environment interactions contribute to a reduction of control effort by performing computation, pre-processing, post-processing, or making computations less costly by providing specialized computing structures. This section presents physical processes that enable behaviour, but that do not contribute to a reduction of the cognitive load. For the development of a unifying perspective, it is important to also understand which type of processes should not be included. Le et al. [106] have investigated the vertical forces exerted by a beetle's elytra (wings). Their numerical simulations show that "*vertical force generated by the elytra without interaction is not sufficient to support even its own weight. However, the elytron-hind wing interaction improves the vertical force on the elytra up to 80%; thus, the total vertical force could fully support its own weight.*" Other studies show that in-phase motions are used during acceleration and forward flight [107] and that they generate larger aerodynamic forces that out-of-phase motions [108]. Here, the interaction of the wings (elytra) with the environment (aerodynamics) clearly contributes to the behaviour. Furthermore, without these interactions, flying would not be possible for the beetle. The question is, do these body-environment interactions reduce the computational burden for the brain? We argue that this is not the case, as there is no way to in-source these body-environment interactions into the brain. The difference to the example given before is that the shaping of the wing described by Wood [63] could, in general, be controlled. Another example are the feet of a gecko, which exploit van der Waals adhesion [109]. Without the exploitation of this effect, the gecko would not be able to walk on vertical surfaces, but this again is an example for body-environment interactions that cannot be replaced by processes in the brain [25].

1.2 What is Morphology, Computation, and Morphological Computation?

The previous sections presented different ways in which a morphology can reduce the computational effort for the brain. This is commonly known as *Morphological Computation*. Although this term is intuitively compelling, it is misleading, which has led to modifications of the way it is used over time.

Initially, *Morphological Computation* was defined more inclusively in the sense that it captured most of the examples that we presented in the previous section. Over time, this has changed to a more technical definition which excludes most examples given above. We will demonstrate that by presenting several quotes several quotes from literature are presented that illustrate how the definition of the term has changed over time. This will be followed by a discussion of different definitions of morphology and computation. We will argue that the term "computation" is the reason why it is understood very narrowly today. This will lead to the proposal of a new definition of morphological contributions to the reduction of computational effort.

1.2.1 How Morphological Computation Has Changed over Time

This section discusses how the understanding and definition of the term *Morphological Computation* has changed over time. For this purpose, we present several quotes from literature in their chronological order. This first quote shows that the concept was initially understood very broadly:

> Having the right morphology to exploit the system-environment interaction nicely illustrates the principle of cheap design. The general point is that often neural processing can be

traded for morphology. The obvious advantages of morphology are speed and little required processing (Pfeifer and Scheier [14]).

Cheap design is here understood as the reduction of control as a result of exploiting the morphology see also [110]. Note, that in this quote it was explicitly discussed that less processing is done by the morphology compared to having the equivalent computation in the brain. Hence, there is not one unit of computation that is divided between the brain to the body. This understanding of morphological computation is also continued in the following two quotes:

> It turns out, however, that quite often morphology can be traded off for computational resources. That is, the choice of an adequate structural setup and of appropriate material properties can substantially simplify the design of the control architecture. It follows that morphology can be conceptualized as implicitly performing some kind of computation. (Matsushita et al. [111])
> It should also be noted that these motor actions are physical processes, not computational ones, but they are computationally relevant, or put differently, relevant for neural processing, which is why we use the term 'morphological computation" (Pfeifer [28]).

The quotes above clearly state that the body is not computing, but the processes are relevant for computation. This is a significant distinction and very close to the definition that will be given for *Morphological Intelligence* below (see Definition 1.1). The next quote is similar in its nature.

In the previous quotes, the body was not computing, whereas in the following quotes this is not as clear any more. Here, morphological computation can be understood as computational processes in the body:

> By 'morphological computation' we mean that certain processes are performed by the body that otherwise would have to be performed by the brain. (Pfeifer et al. [35])
> Morphological Computation refers to the computation which is conducted by the body, that otherwise would have to be performed by the brain. (Pfeifer and Bongard [13])
> Often, morphology and materials can take over some of the functions normally attributed to the brain (or the control), a phenomenon called 'morphological computation". (Pfeifer and Gómez [112])
> A well-designed morphology can ease control requirements by performing some of the computation otherwise required of the controller (Paul [29]).

The following three quotes show very clearly, that in recent years the understanding has changed significantly from the original way it was used. From 2012 on, morphological computation is computation that happens in the body.

> In 2007, a workshop at the first International Conference on Morphological Computing in Venice, Italy, led by Norman Packard, informally defined morphological computing as 'any process that (a) serves for a computational purpose, (b) has clearly assignable input and output states and (c) is programmable, where 'programmable' is understood in the broad sense that a programmer can vary the behaviour of the system by varying a set of parameters." (Füchslin et al. [27])
>
> Morphological computation includes a broad range of different levels of complexity regarding the type of computation (i.e., linear, non-linear, including memory, dynamic, etc.), but also embraces a huge variety of different morphologies (from the molecular level to the large-scale properties of biological organisms). (Hauser et al. [20])

> Morphological computation is a term, which captures conceptually the observation that biological systems take advantage of their morphology to conduct computations needed for successful interaction with their environments (Hauser et al. [113]).

Müller and Hoffmann [25] summarise it, when they say that *"The contribution of the body to cognition and control in natural and artificial agents is increasingly described as 'off-loading computation from the brain to the body', where the body is said to perform 'morphological computation'."* They also state that the off-loading perspective is misleading because *"the contribution of body morphology to cognition and control is rarely computational."* We will revisit this point in Sect. 1.2.

We believe that one of the reasons for this shift in perspective from computationally relevant processes to actual physical computation results from the term computation. Hence, it seems helpful to look at how the terms "morphology" and "computation" are used in the literature.

1.2.2 Definition of the Term Morphology

The word *morphology* originates from the Greek word *morphē* (form) and the English suffix -*logy* (study of). In biology, morphology refers to the study of the form of living organisms. The meaning has not changed significantly in the context of embodied artificial intelligence. Paul [29] defines morphology very similar to the way it is used in biology. She defines the morphology of a robot as its physical structure. This includes the specific characteristics of the body's parts, e.g. link sizes, number of links, joint characteristics, mass distribution, actuator characteristics, material properties, sensor characteristics and sensor placements. She summarises it as *"any characteristic which defines the physical structure of the robot is included in the term morphology."*

Very similarly, Zambrano et al. [15] define morphology as *"the shape, the geometry, the placement and the compliance properties of the body parts define the perception and the interaction with the environment, thus connecting such kind of features with the expressed behaviour, synergistically."* The new aspect here, compared to the definition given by Paul [29], is the behaviour relevance of the body's physical properties. The definition of Zambrano et al. [15] includes the definition of Paul [29] and is more suitable with respect to this work. We are not interested in, e.g. the colour of a soft manipulator as long as it does not influence the hand's behaviour. Hence, in the remainder of this work, whenever we use the terms *morphology, physical properties of the body* or *physical properties of the morphology*, we refer to the physical characteristics of a body that are behaviour relevant in the sense of Zambrano et al. [15].

1.2.3 Definition of the Term Computation

The theory of computation is most often related to the Turing machine, which Turing introduced as the *a*-machine in his famous work titled "*On Computable Numbers, with an Application to the Entscheidungsproblem*" [114]. Generally speaking, a Turing machine is an infinite tape, a head that can read and write from the tape, a finite set of symbols, and a finite set of instruction (move left, move right, etc.). Hopcroft et al. [115] describe a Turing machine formally as $M = (Q, \Sigma, \Gamma, \delta, q_0, B, F)$, where Q is a finite set of states, Σ is a finite set of input symbols, Γ is the complete set of tape symbols, i.e., $\Sigma \subset \Gamma$, δ is a transition function (table of instructions), q_0 is the initial state of the tape, i.e., $q_0 \in Q$, B is the blank symbol, which is in Γ but not part of Σ, and finally, F is the set of final states ($B \subset Q$), for which the process is halted.

The significance of the Turing machine can be summarised in the following statement. Anything that is computable is computable with the Turing machine [114]. In the context of computer science, computation has become synonymous with Turing completeness. This means that a device is a computer only if it is as powerful as a Turing machine. Following this argumentation, morphological computation would mean that a morphology is Turing complete. This leads to the philosophical concept of pancomputationalism [116], which is discussed next.

In general, pancomputationalism refers to the notion that every physical system is a computer. Varieties of pancomputationalism differ in how much computation is attributed to any system [116]. The strongest form of pancomputationalism claims that every physical system can compute any kind of computation [117]. Putnam [118] gives the following illustration. The dynamics of any physical system in continuous time can be sliced into a time series of discrete states. The resulting states are then aggregated so that they correspond to an arbitrary sequence of computational states. In other terms, one can think of the discrete states as bits and the aggregated states as bytes, which can then represent any sequence of numbers. Putman argues that this way, every physical system implements every finite-state machine (cited from [116]). This is very close to the version of pancomputationalism that is discussed in [25]. In this version of pancomputationalism, the entire universe is considered to be a computer in the literal sense. Morphological computation is then the amount of computation that is not located in the brain. We will discuss this point of view in more detail below (see Sect. 1.2).

Horsman et al. [119] offer a different perspective on computation. The motivation for their work is to understand what differentiates a computing device from a purely physical system, e.g. what differentiates a pocket calculator from a river. The authors identify four fundamental properties that any computing system must fulfil. We will briefly summarise them first, and elaborate on them afterwards. *First*, there needs to be an established theory of the physical computation device. *Second*, there must be an encoding and decoding function to encode abstract entities into physical representations and to decode the physical state of the machine back into abstract entities. *Third*, there must be at least one fundamental physical process that takes the

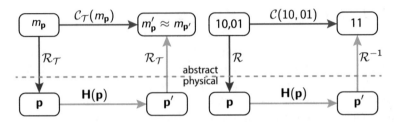

Fig. 1.6 *Commuting diagram of physical computation.* This diagram shows how computation is understood by Horsman et al. [119]. On the left-hand side, we have the general commutative diagram, where m_p is the abstract entity that is encoded by $\mathcal{R}_T$ into the physical system resulting in the state **p**. The physical computation is done by fundamental physical processes, denoted by $\mathbf{H(p)}$ and results in the final physical state $\mathbf{p}'$. This state is decoded by $\mathcal{R}_T$ back into the abstract entity $m_{p'}$. The system is said to be computing if the decoded state $m_{p'}$ corresponds with the prediction m'_p that is generated by the theory $\mathcal{C}_T(m_p)$. The figure on the right-hand side shows the example of the computation $2 + 1 = 3$, represented by bits

input and transforms it into the output. Finally, *fourth*, there must be a commuting diagram (see Fig. 1.6), that explains the overall process.

To explain the four properties, we use a pocket calculator and the calculation $2 + 2 = 4$ as an example. The first property, the established theory, means that the physical properties that operate in the calculator must be understood, and more importantly, predictable. In other words, a random physical process is not computing. In the example of the pocket calculator, we have a good understanding of how e.g. the transistors that perform the calculation work. The second property, encoding and decoding, means that there must be a function that takes the abstract entity, e.g. the number 2, and encodes it into a physical state of the machine. In this case, it might be the binary representation of a byte, i.e., 00000010, which is encoded by different voltage levels in the memory of the pocket calculator. The decoding function means that the result of our calculation, the voltages that represent the byte 00000100, will be decoded into the number 4 on the display of the calculator. One important aspect is hidden in the second property. Encoding and decoding mean that there must be an encoder and decoder, i.e., somebody who sets up the physical computer (encoding) and reads out its final state (decoder). As Müller and Hoffmann [25] nicely summarise the work by [119]; without the user, there is no computer. This does not necessarily have to be a human, but without the decoding step, i.e., reading of the output and translating it back into an abstract representation, Horsman et al. [119] argue, computation becomes a meaningless physical process. In essence, encoding and decoding distinguish the pocket calculator from the river. The third property, fundamental physical process, relates to electricity running through the circuits of our calculator changing the states of transistors. The final property, the commuting diagram (see Fig. 1.6) means that our theory or prediction of the calculations must match the results of the encoding, physical process, and decoding.

Now that we have defined morphology and computation, we can investigate, how the different types of morphological contributions to intelligence presented in the previous sections fit within these definitions.

1.2.4 What is Morphological Computation?

From the definitions given in the previous section, it does not seem that the notion of morphology is problematic in the context of this work. In general, it captures behaviour relevant aspects of the morphology, e.g., sensors and their placement, friction, segment shapes and placement, etc. In all the examples given in this chapter (see Sect. 1.1), this definition fits very well. The problematic part of the term *morphological computation* results from the word *computation*. In the previous section, three definitions were given, i.e., Turing-completeness, pancomputationalism, and the theory of physical computation by Horsman et al. [119]. Requiring that morphologies are Turing-complete would exclude all previously mentioned examples from the notion of morphological computation unless they are understood in the context of pancomputationalism. This means that Turing completeness is only applicable, if either every physical process is understood as computation in the way Putnam [118] describes it or if the universe is a computer whose computation is distributed between its components, of which the two of interest are the system's brain and everything else. This will be discussed in the next paragraphs.

We will revisit the examples presented in the previous sections in the context of the definitions given above. One of the examples was the Passive Dynamic Walker [8], which is often cited as an instance of morphological computation (see e.g. [13, 15, 20, 21, 112]). Yet others conclude that Passive Dynamic Walking should be understood only as a passive mechanical process [120, 121]. Müller and Hoffmann [25] question the notion of computation with respect to the Passive Dynamic Walker and go so far to say that if the Passive Dynamic Walker, a purely mechanical system, is considered to perform computation, then there is no system that cannot be considered to perform computation. In their words, *"cognitive and morphological processes are trivially computational in the sense that everything else is"*, which is also known as (limited) pancomputationalism [122] (see the previous section). To conclude this example, it seems that there is the desire to attribute morphological computation to the Passive Dynamic Walker, but that there are good arguments to refute this position. The strongest argument is that it is a purely physical system, and in this form, equivalent to a ball rolling downhill. If a ball rolling downhill is performing morphological computation, then everything is performing morphological computation, which make the concept useless.

On the other hand, there is a consensus that morphological computation is equivalent to physical reservoir computing [21, 25, 27]. This was addressed earlier in Sect. 1.1.1 and will also be discussed in more detail in the next chapter (see Chap. 3). Reservoir computing fits well with the notion of physical computation provided by Horsman et al. [119] for the following reason. A reservoir computer [38, 39, 123] is a recurrent neural network whose internal dynamics can be influenced by an input layer and are read by an output layer. In terms of the theory by Horsman et al. [119] there is an encoding (input layer, $\mathcal{R}$, see Fig. 1.6), a physical dynamics that transforms the input to an output ($\mathbf{H(p)}$, see Fig. 1.6), and a decoding function (output layer, $\mathcal{R}^{-1}$, see Fig. 1.6). Furthermore, we have a theory of the system's progression over time ($\mathcal{C}$, see Fig. 1.6). Research into physical reservoir computing has resulted in a variety of interesting examples of physical implementations of analogue processors. Physical reservoir computing has been shown to work with water ripples [124], mechanical constructs [21, 66], sound waves [125], electro-optical devices [126, 127], fully optical devices [128] and nanophotonic circuits [129, 130].

At first, it seems that physical reservoir computing is a non-controversial interpretation of morphological computation. But this does not hold if this concept is investigated in more detail. Under the definition given by Horsman et al. [119], the spine-drive robot (see Sect. 1.1.2 and [51]) clearly is an instance of a physical computer. We have an encoding and decoding process, and hence, also an encoder and decoder (the controller). There is a physical process that transforms inputs to outputs (physical reservoir computing) and a theory of how this is done (mechanics). The ripples in a bucket of water [124] are equivalent, because there are motors, which represent the encoding function and a digital camera which represents the decoding function. A theory of how the waves propagate in the bucket exists. In this context, the Passive Dynamic Walker must also be considered as a physical computer for the following reason. The initial state of the Passive Walker is carefully controlled by a human. The amount of initial deviation and initial force applied to the swinging leg is crucial for an emerging walking pattern. Hence, encoding is present in this system. The same holds for decoding, as the system is carefully observed and the walking is monitored. Passive walkers are usually evaluated by the number of steps they can perform before falling over. The underlying physics are well-understood and mathematical models for their simulation exist. Hence, the Passive Dynamic Walker as a mechanical system fulfils the criteria set forth by Horsman et al. [119]. We can now also understand the difference between the Passive Dynamic Walker and the ball rolling downhill. In very simplistic terms, if nobody is watching the ball rolling downhill, it is not computing. If someone places the ball on a slope and uses the time it requires to travel a certain distance to calculate e.g. the inclination, then the ball becomes a physical computer. [119] state that the distinction between a purely physical system and a physical computer relies on the interpretation of an external observer. Although this is a valid point, it does not seem very useful in determining if a system is performing morphological computation. Any physical process can be said to perform computation under this definition, even the gecko's feet, as the gecko clearly encodes the state of the feet by placing them and reads their state to decide if it is safe to perform the next step. The gecko knows what will happen, if it places its feet on a surface, hence, one can argue that it has some kind of theory of it. In summary, the commuting diagram shown in Fig. 1.6 can also be drawn for gecko walking up a wall, and hence, for almost any process in a cognitive system.

This discussion shows that the term *computation* is misleading and results in more confusion than clarification about morphological contributions to the reduction of the cognitive load. As a possible solution to this problem, Müller and Hoffmann [25] suggest to distinguish between three different cases, namely *"(1) morphology that facilitates control, (2) morphology that facilitates perception and the rare cases of (3) proper morphological computation, such as reservoir computing."* As we have discussed in the previous sections, this already captures many different aspects of morphological contributions to behaviour, but not all. The other problem we see is that this leads to diversification and possibly also trivialisation of this central concept in the field of embodied artificial intelligence. If we follow this line of argumentation, we have plausibly reduced morphological computation to systems what can clearly be related to physical reservoir computing. As discussed by Müller and Hoffmann [25], this leaves us with a small subset of interesting systems, or as we have argued, can be applied to any system with Horsman's theory of physical computation.

We believe that this is the wrong approach. Instead of diversifying morphological contributions to intelligence, we should find a unifying perspective, which is the goal of this work. A proposal for a unifying perspective is discussed in the next section.

1.3 Morphological Intelligence

The previous sections made the case that the concept of a morphology that is computing is problematic in the sense that is has resulted in a more and more technical interpretation of the notion of *Morphological Computation* up to the point that it is now only applicable to systems which are closely related to physical reservoir computing. We believe that instead of narrowing the set of systems that are of interest within this concept, we should find a conceptual umbrella that captures all possible contribution of morphology to the reduction of computation in the brain. We propose the new term *Morphological Intelligence*. This section will explain why this is not just exchanging one problematic term by another, but why this is a valuable conceptual shift.

First, the term intelligence has no clear definition. Which seems to be a disadvantage is an advantage. Replacing *computation* with *intelligence* captures the idea that the morphology is important with respect to intelligence, but without the burden of a narrow definition of how this is exactly done. This omits the problem of having to fit observations with clear conceptual definitions as it is the case with the term *computation*. To understand why we argue for the term *Morphological Intelligence*, we will give a conceptual definition of intelligence, that is not tailored to morphology but applicable in a larger context.

In this book, we follow Krakauer's definition of intelligence and stupidity [131, 132]. The idea is illustrated using the Rubik's cube, which is explained next. A person is handed a Rubik's cube that is in a randomised initial state and asked to perform random actions, i.e., random rotations on the cube. Eventually, when enough time has passed, the Rubik's cube will be solved, meaning that each side will only have patches of the same colour. This random behaviour, which does not is able to solve the Rubik's cube. It is important to note here, that his strategy does not require any memory or strategy. If we now hand a randomly initialised Rubik's to an expert, she or he will be able to solve this cube in less than 25 moves [133]. This requires a lot of training and the sophistication of a solving strategy which includes solving several steps at once (for an overview of speedsolving methods, see [134]). In this case, memory, planning, strategy, etc. come into play which allows a trained person to solve the cube significantly faster (in time and number of steps) compared to a random behaviour (or even an untrained person). Such behaviour can be said to be *intelligent*. On the other hand, if a person would always choose the same motion, e.g. always rotate only one side of the cube, he or she would not solve the Rubik's cube. Such a behaviour is called *stupid* by Krakauer. To summarise, any behaviour that is better than a baseline behaviour is considered intelligent and any behaviour that is worse than a baseline behaviour is considered stupid. Krakauer nicely states this as "*Intelligence is making a hard problem easy*" and "*Stupidity is making an easy problem difficult.*"

Adapting this to the context of embodied artificial intelligence, we state that a morphology should be called intelligent, if it makes the task of control easier, i.e., if it requires less control effort compared to some baseline controller that does not exploit the morphology. In the same way, a morphology could be called *stupid*, if it makes the task of control more difficult. A suggestion of how to quantify this will be given in the next chapter (see Sect. 3.7).

Hence, we give the following definition for *Morphological Intelligence*:

Definition 1.1 (*Morphological Intelligence*) Morphological Intelligence is the reduction of computational cost for the brain (or controller) resulting from the exploitation of the morphology and its interaction with the environment.

Consequently, *Morphological Stupidity* is the increase of computational cost resulting from the morphology and its interaction with the environment.

In the following paragraphs, we will explain the definition of *Morphological Intelligence* in more detail and why we believe that it captures all of the previously discussed examples. For this purpose, we will also contrast the definition of *Morphological Intelligence* with the most general and most commonly used definition of *Morphological Computation* given by Pfeifer and Bongard [13], which is "*Morphological Computation refers to the computation which is conducted by the body, that otherwise would have to be performed by the brain.*"

The first difference is that the new definition does not require that any computation is actually performed by the body. It defines *Morphological Intelligence* as the reduction of computational cost in the brain resulting from body-environment interactions. We believe that this is a valuable conceptual change as it omits the problem of finding the right definition of computation that can be applied to physical processes. Furthermore, using *computational cost* instead of computation is again less narrow. The term *computational cost* can refer to the amount of computation, e.g., time step cost [135] or the number of hidden neurons in a specific model [110], but it can also refer to energetic cost, as e.g. described by Clark and Sokoloff [136]. In most cases, it will refer to the cost in terms of computational complexity. The last modification with respect to the definition given by Pfeifer and Bongard [13] is that we not only include processes that happen within the body, but also its interaction with the environment.

We believe that this definition captures all of the previously mentioned morphological contributions to a behaviour (see above) as well as it excludes those that should not be included for the following reasons. Morphological computation, as described in Sect. 1.1.1 is included, as it is the computation performed by the body that does not have to be performed by the brain, and hence, it is the reduction of computational cost. Morphological control was defined as a controller that uses the results of morphological control to generate motor commands. This would also be captured in the new definition as the controller itself is reduced in its complexity as part of the computation that is outsourced to the body. Pre- and post-processing (see Sects. 1.1.3 and 1.1.4) are also processes which are captured by the new definition, because the pre-processing that is performed by sensors (examples were given for the eye and ear) and the post-processing that is performed by actuators (examples were given for muscles and tensegrity robots) reduce the controller complexity. Note, that we no longer have to be concerned if these processes, e.g. in the muscle (see Sect. 1.1.4) are well-understood as computation. Brain, or more generally, computing layouts are also physical properties of the morphology that reduce the computation cost. An example of the L2 cache and the wiring of the superior colliculus (gaze control network) was given in Sect. 1.1.5. Finally, the gecko's feet are not considered to increase *Morphological Intelligence* as the van der Waals adhesion cannot be compensated with an increase in computational complexity.

To summarise, we believe that the definition of *Morphological Intelligence* as the body making the difficult task of control easier is more general and inclusive with respect of the various form in which a morphology and its interaction with its environment can contribute to cognition and intelligence. Furthermore, it omits the problem of finding the right definition of computation that can be applied to physical processes, which so far has not been a very fruitful discussion in this context.

1.4 Organisation of This Book and Main Results

This book is organised in the following way.

Chapter 2 presents the formal basis for this book. Since almost all presented quantifications are based on information theory, the second chapter discusses the fundamental concepts, i.e., entropy, relative entropy (Kullback-Leibler divergence), mutual information, etc. Important theorems and their proofs are provided. Up to this point, the discussion will have taken place only for discrete systems, i.e., random variables with a finite alphabet. The goal of the work presented in this book is to apply the proposed concepts to real-world data, which is why the second chapter closes with an overview of current non-parametric estimations of entropy, mutual information, and conditional mutual information on continuous data.

Chapter 3 presents the main results of this book. It starts with an overview of related work on formalising morphological computation. These are formulated in the context of dynamical systems and reservoir computing. Related work in the context of information theory is also presented, although these focus on quantifying control effort and are only indirectly related to quantifying morphological intelligence. After the presentation of related work, five different concepts of quantifying morphological intelligence are discussed. The first three build upon our previously published work. The last two concepts are new had have not been published elsewhere. The presentation of the measures resembles the evolution of the theory of morphological computation. This chapter is the first comprehensive overview of the different quantifications of morphological intelligence. Relations between the measures are drawn and discussed.

Chapter 4 presents a numerical analysis of the measures presented in the third chapter. This chapter is important because it highlights the common features and differences of the measures, that are not easily identifiable from an analytic approach alone. Similar experiments to those presented in chapter four have been conducted for the measures MI_W, MI_A, $UI(W' : W \setminus A)$, $CI(W' : W; A)$, and MI_{SY}, but not to the detail presented here. This is important to point out because the chapter presents new findings for $UI(W' : W \setminus A)$ and $CI(W' : W; A)$, which were not seen previously. This is the first comprehensive overview of the numerical properties of the proposed measures. The results presented in the fourth chapter lead to new insights, which measures should be used in real-world applications.

Chapter 5 presents two applications of two measures, MI_W and MI_{MI}, in two fields of interest, namely soft robotics and biomechanics. This chapter builds upon previously published results but is adapted in the light of the recent developments presented in this book. The goal the work presented in the fifth chapter was to distinguish body-environment interactions which support the task (morphological intelligence) from those which make the task more difficult (morphological stupidity) and to make the results of the analysis operational for an automated design process of soft robots. In the second experiment, we analyse two different muscle models and compare their performance with respect to morphological intelligence with the performance of a DC-motor. The results show that a non-linear muscle contributes significantly more to the hopping behaviour compared to the other two models. Analysis of morphological intelligence at every point in time reveals that the non-linear muscles model contributes significantly more to the behaviour compared to the flight phase, i.e., the phase in which the hopper does not touch the ground. This is important because, during the flight phase, the hopper's behaviour only results from the interaction of the hopper with gravity.

Chapter 6 closes this book with a discussion.

The appendix introduces *gomi*, which is a software package to calculate Morphological Intelligence on data.

References

1. Brooks RA (1986) A robust layered control system for a mobile robot. IEEE J Robot Autom 2(1):14–23
2. Brooks RA (1991a) Intelligence without reason. In: Myopoulos J, Reiter R (eds) Proceedings of the 12th international joint conference on artificial intelligence (IJCAI-91), Morgan Kaufmann publishers Inc.: San Mateo, CA, USA, Sydney, Australia, pp 569–595
3. Brooks RA (1991b) Intelligence without representation. Artif Intell 47(1–3):139–159
4. Haugeland J (1985) Artificial intelligence: the very idea. MIT Press, Cambridge, MA
5. Siciliano B, Khatib O (eds) (2016) Springer handbook of robotics. Springer, Berlin, Heidelberg
6. Honda (2018) Asimo. http://asimo.honda.com/news/
7. McGeer T (1990b) Passive walking with knees. In: Robotics and automation, pp 1640–1645
8. McGeer T (1990a) Passive dynamic walking. Int J Robot Res 9(2):62–82
9. Hirose M, Haikawa Y, Takenaka T, K H (2001) Development of humanoid robot asimo. In: Proceedings IEEE/RSJ international conference on intelligent robots and systems (Oct 29, 2001)
10. Vukobratović M, Stepanenko J (1972) On the stability of anthropomorphic systems. Math Biosci 15(1):1–37
11. Collins SH (2017) Passive-dynamic walking robot—photos. https://www.andrew.cmu.edu/user/shc17/Passive_Robot/PassiveRobot_photos.htm
12. Collins S, Ruina A, Tedrake R, Wisse M (2005) Efficient bipedal robots based on passive-dynamic walkers. Science 307(5712):1082–1085
13. Pfeifer R, Bongard JC (2006) How the body shapes the way we think: a new view of intelligence. The MIT Press (Bradford Books), Cambridge, MA
14. Pfeifer R, Scheier C (1999) Understanding intelligence. MIT Press, Cambridge, MA, USA
15. Zambrano D, Cianchetti M, Laschi C, Hauser H, Füchslin R, Pfeifer R (2014) The observation of the morphological computation phenomenon in nature is the first step for the formalization of the principle. Opinions and Outlooks on Morphological Computation pp 214–225
16. Rückert EA, Neumann G (2013) Stochastic optimal control methods for investigating the power of morphological computation. Artif Life 19(1):115–131
17. Reis M, Yu X, Maheshwari N, Iida F (2012) Morphological computation of multi-gaited robot locomotion based on free vibration. Artif Life 19(1):97–114
18. Pfeifer R, Iida F (2005) Morphological computation: connecting body, brain and environment. Japanese Sci Mon 58(2):48–54
19. Nakajima K, Hauser H, Kang R, Guglielmino E, Caldwell D, Pfeifer R (2013a) A soft body as a reservoir: case studies in a dynamic model of octopus-inspired soft robotic arm. Front Comput Neurosci 7:91
20. Hauser H, Ijspeert AJ, Füchslin RM, Pfeifer R, Maass W (2012) The role of feedback in morphological computation with compliant bodies. Biol Cybern 106(10):595–613
21. Hauser H, Ijspeert A, Füchslin RM, Pfeifer R, Maass W (2011) Towards a theoretical foundation for morphological computation with compliant bodies. Biol Cybern 105(5–6):355–370
22. Pfeifer R, Gómez G (2009) Creating brain-like intelligence. In: Sendhoff B, Körner E, Sporns O, Ritter H, Doya K (eds) Creating brain-like intelligence: from basic principles to complex intelligent systems, Springer, Berlin, Heidelberg, chap Morphological Computation—Connecting Brain, Body, and Environment, pp 66–83
23. Lichtensteiger L (2004) The need to adapt and its implications for embodiment. Springer, Berlin, Heidelberg, pp 98–106
24. Nowakowski PR (2017) Bodily processing: the role of morphological computation. Entropy 19(295):
25. Müller VC, Hoffmann M (2017) What is morphological computation? on how the body contributes to cognition and control. Artif Life 23(1):1–24
26. Hoffmann M, Müller VC (2017) Simple or complex bodies? trade-offs in exploiting body morphology for control. Springer International Publishing, Cham, pp 335–345

27. Füchslin RM, Dzyakanchuk A, Flumini D, Hauser H, Hunt KJ, Luchsinger RH, Reller B, Scheidegger S, Walker R (2012) Morphological computation and morphological control: Steps toward a formal theory and applications. Artif Life 19(1):9–34

28. Pfeifer R, Iida F, Gòmez G (2006) Morphological computation for adaptive behavior and cognition. Int Congr Ser 1291:22–29

29. Paul C (2006) Morphological computation: a basis for the analysis of morphology and control requirements. Robot Auton Syst 54(8):619–630

30. Iida F, Pfeifer R (2006) Sensing through body dynamics. Robot Auton Syst 54(8):631–640

31. Iida F, Pfeifer R (2004) "cheap" rapid locomotion of a quadruped robot: self-stabilization of bounding gait. In: Proceedings of the international conference on intelligent autonomous systems, pp 642–649

32. Iida F, Gomez G, Pfeifer R (2005) Exploiting body dynamics for controlling a running quadruped robot. In: ICAR '05 Proceedings, 12th international conference on advanced robotics, pp 229–235

33. Iida F (2005) Cheap design and behavioral diversity for autonomous adaptive robots. PhD thesis, University of Zurich

34. Lungarella M, Pegors T, Bulwinkle D, Sporns O (2005a) Methods for quantifying the informational structure of sensory and motor data. Neuroinformatics 3(3):243–262

35. Pfeifer R, Lungarella M, Iida F (2007) Self-organization, embodiment, and biologically inspired robotics. Science 318(5853):1088–1093

36. Iida F, Pfeifer R (2005) Structuring sensory information through body dynamics. In: IROS05 Workshop on Morphology, Control and Passive Dynamics, http://people.csail.mit.edu/iida/papers/iida_iros05ws_cr.pdf

37. Jaeger H (2002b) Tutorial on training recurrent neural networks, covering BPPT, RTRL, EKF and the "echo state network" approach, vol 5. GMD-Forschungszentrum Informationstechnik

38. Jaeger H, Haas H (2004) Harnessing nonlinearity: predicting chaotic systems and saving energy in wireless communication. Science 304(5667):78–80

39. Maass W, Natschläger T, Markram H (2002) Real-time computing without stable states: a new framework for neural computation based on perturbations. Neural Comput 14(11):2531–2560

40. Buonomano DV, Maass W (2009) State-dependent computations: spatiotemporal processing in cortical networks. Nat Rev Neurosci 10(2):113–125

41. Verstraeten D, Schrauwen B, d'Haene M, Stroobandt D (2007) An experimental unification of reservoir computing methods. Neural Netw 20(3):391–403

42. Nakajima K, Hauser H, Kang R, Guglielmino E, Caldwell DG, Pfeifer R (2013b) Computing with a muscular-hydrostat system. In: 2013 IEEE international conference on robotics and automation, pp 1504–1511

43. Nakajima K, Li T, Hauser H, Pfeifer R (2014) Exploiting short-term memory in soft body dynamics as a computational resource. J Royal Soc Interface 11(100):

44. Nakajima K, Hauser H, Li T, Pfeifer R (2015) Information processing via physical soft body. Sci Reports 5:10487 EP

45. Fausch KD (1984) Profitable stream positions for salmonids: relating specific growth rate to net energy gain. Can J Zool 62(3):441–451

46. Beal DN, Hover FS, Triantafyllou MS, Liao JC, Lauder GV (2006) Passive propulsion in vortex wakes. J Fluid Mech 549:385–402

47. Liao JC, Beal DN, Lauder GV, Triantafyllou MS (2003b) The kármán gait: novel body kinematics of rainbow trout swimming in a vortex street. J Exp Biol 206(6):1059–1073

48. Liao JC, Beal DN, Lauder GV, Triantafyllou MS (2003a) Fish exploiting vortices decrease muscle activity. Science 302(5650):1566–1569

49. Wu TY (1972) Extraction of flow energy by a wing oscillating in waves. J Ship Res pp 66–78

50. Wu TY, Chwang AT (1975) Extraction of flow energy by fish and birds in a wavy stream. Springer, US, Boston, MA, pp 687–702

51. Zhao Q, Nakajima K, Sumioka H, Hauser H, Pfeifer R (2013) Spine dynamics as a computational resource in spine-driven quadruped locomotion. In: 2013 IEEE/RSJ international conference on intelligent robots and systems, pp 1445–1451

52. McEvoy MA, Correll N (2015) Materials that couple sensing, actuation, computation, and communication. Science 347(6228):

53. Franceschini N, Pichon JM, Blanes C, Brady J (1992) From insect vision to robot vision [and discussion]. Philos Trans Royal Soc B: Biol Sci 337(1281):283–294

54. Carvell G, Simons D (1990) Biometric analyses of vibrissal tactile discrimination in the rat. J Neurosci 10(8):2638–2648

55. Waiblinger C, Brugger D, Whitmire CJ, Stanley GB, Schwarz C (2015) Support for the slip hypothesis from whisker-related tactile perception of rats in a noisy environment. Front Integr Neurosci 9:53

56. Lucianna FA, Farfán FD, Pizá GA, Albarracín AL, Felice CJ (2016) Functional specificity of rat vibrissal primary afferents. Physiol Reports 4(11):

57. Carvell GE, Simons DJ (2017) Effect of whisker geometry on contact force produced by vibrissae moving at different velocities. J Neurophysiol 118(3):1637–1649

58. Georgieva P, Brugger D, Schwarz C (2014) Are spatial frequency cues used for whisker-based active discrimination? Front Behav Neurosci 8:379

59. Fend M, Bovet S, Yokoi H, Pfeifer R (2003) An active artificial whisker array for texture discrimination. In: Proceedings 2003 IEEE/RSJ international conference on intelligent robots and systems (IROS 2003) (Cat. No.03CH37453), vol 2, pp 1044–1049

60. Lungarella M, Hafner VV, Pfeifer R, Yokoi H (2002) An artificial whisker sensor for robotics. In: IEEE/RSJ international conference on Intelligent robots and systems, IEEE, vol 3, pp 2931–2936

61. Hosoda K (2004) Robot finger design for developmental tactile interaction. Springer, Berlin, Heidelberg, pp 219–230

62. Wootton RJ (1992) Functional morphology of insect wings. Ann Rev Entomol 37(1):113–140

63. Wood RJ (2007) Design, fabrication, and analysis of a 3dof, 3cm flapping-wing MAV. (2007) IEEE/RSJ International Conference on Intelligent Robots and Systems, October 29-November 2, 2007. Sheraton Hotel and Marina, San Diego, California, USA, pp 1576–1581

64. Fuller RB (1961) Tensegrity. Portfolio and Art News Annua 4

65. Fuller RB (1962) Tensile-integrity structures. United States Patent 3(063):521

66. Caluwaerts K, D'Haene M, Verstraeten D, Schrauwen B (2012) Locomotion without a brain: physical reservoir computing in tensegrity structures. Artif Life 19(1):35–66

67. Caluwaerts K, Despraz J, Işçen A, Sabelhaus AP, Bruce J, Schrauwen B, SunSpiral V (2014) Design and control of compliant tensegrity robots through simulation and hardware validation. J Royal Soc Interface 11(98):

68. Caluwaerts K, Schrauwen B (2011) The body as a reservoir: locomotion and sensing with linear feedback. In: Conference proceedings: 2nd international conference on morphological computation, p 3

69. Paul C, Roberts JW, Lipson H, Cuevas FJV (2005) Gait production in a tensegrity based robot. In: ICAR '05. Proceedings, 12th international conference on advanced robotics, pp 216–222

70. Rieffel JA, Valero-Cuevas FJ, Lipson H (2010) Morphological communication: exploiting coupled dynamics in a complex mechanical structure to achieve locomotion. J Royal Soc Interface 7(45):613–621

71. Agogino A, SunSpiral V, Atkinson D (2013) Super ball bot—structures for planetary landing and exploration. NASA Innovative Advanced Concepts (NIAC) Program, Phase 1, Final Report

72. SunSpiral V, Agogino A, , Atkinson D (2015) Super ball bot—structures for planetary landing and exploration. NASA Innovative Advanced Concepts (NIAC) Program, Phase 2, Final Report

73. NASA (2017) Superball bot tensegrity planetary lander. https://ti.arc.nasa.gov/tech/asr/intelligent-robotics/tensegrity/superballbot/

74. Calladine C (1978) Buckminster fuller's "tensegrity" structures and clerk maxwell's rules for the construction of stiff frames. Int J Solids Struct 14(2):161–172

75. Toth TI, Grabowska M, Schmidt J, Büschges A, Daun-Gruhn S (2013) A neuro-mechanical model explaining the physiological role of fast and slow muscle fibres at stop and start of stepping of an insect leg. PLOS ONE 8(11):1–14

76. Blümel M, Guschlbauer C, Daun-Gruhn S, Hooper SL, Büschges A (2012) Hill-type muscle model parameters determined from experiments on single muscles show large animal-to-animal variation. Biol Cybern 106(10):559–571

77. Haeufle DFB, Grimmer S, Kalveram KT, Seyfarth A (2012) Integration of intrinsic muscle properties, feed-forward and feedback signals for generating and stabilizing hopping. J Royal Soc Interface 9(72):1458–1469

78. Haeufle DFB, Grimmer S, Seyfarth A (2010) The role of intrinsic muscle properties for stable hopping-stability is achieved by the force-velocity relation. Bioinspiration Biomim 5(1):016004

79. Haeufle DFB, Günther M, Wunner G, Schmitt S (2014) Quantifying control effort of biological and technical movements: an information-entropy-based approach. Phys Rev E 89:012716

80. Ghazi-Zahedi K, Haeufle DF, Montufar GF, Schmitt S, Ay N (2016) Evaluating morphological computation in muscle and dc-motor driven models of hopping movements. Front Robot AI 3(42)

81. Zhang Z, Yang J, Yu H (2014) Effect of flexible back on energy absorption during landing in cats: a biomechanical investigation. J Bionic Eng 11(4):506–516

82. Hildebrand M (1977) Analysis of asymmetrical gaits. J Mamm 58(2):131

83. Alexander RM, Langman VA, Jayes AS (1977) Fast locomotion of some african ungulates. J Zool 183(3):291–300

84. Hildebrand M (1989) The quadrupedal gaits of vertebratesthe timing of leg movements relates to balance, body shape, agility, speed, and energy expenditure. BioScience 39(11):766

85. Smith JL, Chung SH, Zernicke RF (1993) Gait-related motor patterns and hindlimb kinetics for the cat trot and gallop. Exp Brain Res 94(2):308–322

86. Bertram JEA, Gutmann A (2009) Motions of the running horse and cheetah revisited: fundamental mechanics of the transverse and rotary gallop. J R Soc Interface 6(35):549–59

87. Hudson PE, Corr SA, Wilson AM (2012) High speed galloping in the cheetah (acinonyx jubatus) and the racing greyhound (canis familiaris): spatio-temporal and kinetic characteristics. J Exp Biol 215(14):2425–2434

88. Wilson AM, Lowe JC, Roskilly K, Hudson PE, Golabek KA, McNutt JW (2013) Locomotion dynamics of hunting in wild cheetahs. Nature 498(7453):185–189

89. English AW (1980) The functions of the lumbar spine during stepping in the cat. J Morphol 165(1):55–66

90. Hackert R, Schilling N, Fischer MS (2006) Mechanical self-stabilization, a working hypothesis for the study of the evolution of body proportions in terrestrial mammals? Compt Rendus Palevol 5(3–4):541–549

91. Maes LD, Herbin M, Hackert R, Bels VL, Abourachid A (2007) Steady locomotion in dogs: temporal and associated spatial coordination patterns and the effect of speed. J Exp Biol 211(1):138–149

92. Koob TJ, Long JH Jr (2000) The vertebrate body axis: Evolution and mechanical function1. Am Zool 40(1):1

93. Alexander RM, Jayes AS (1981) Estimates of the bending moments exerted by the lumbar and abdominal muscles of some mammals. J Zool 194(3):291–304

94. Alexander RM, Dimery NJ, Ker RF (1985) Elastic structures in the back and their rôle in galloping in some mammals. J Zool 207(4):467–482

95. Alexander RM (1988) Why mammals gallop. Am Zool 28(1):237–245

96. Fischer MS (1994) Crouched posture and high fulcrum, a principle in the locomotion of small mammals: The example of the rock hyrax (procavia capensis) (mammalia: Hyracoidea). J Human Evolut 26(5):501–524

97. Schilling N, Hackert R (2006) Sagittal spine movements of small therian mammals during asymmetrical gaits. J Exp Biol 209(19):3925–3939

98. Biancardi CM, Minetti AE (2012) Biomechanical determinants of transverse and rotary gallop in cursorial mammals. J Exp Biol 215(23):4144–4156

99. Dermitzakis K, Morales MR, Schweizer A (2012) Modeling the frictional interaction in the tendon-pulley system of the human finger for use in robotics. Artif Life 19(1):149–169

100. Quinn TH, Baumel JJ (1993) Chiropteran tendon locking mechanism. J Morphol 216(2):197–208

101. Spitzenberger F, Eberl-Rothe G (1974) Der Sohlenhaftmechanismus von Dryomys laniger. I. Teil: Makroskopische Untersuchung. II. Teil: Mikroskopische Untersuchung. Annalen des Naturhistorischen Museums in Wien 78:485–494

102. Schutt WAJ (1993) Digital morphology in the chiroptera: the passive digital lock. Acta Anat (Basel) 148(4):219–227

103. Tabareau N, Bennequin D, Berthoz A, Slotine JJ, Girard B (2007) Geometry of the superior colliculus mapping and efficient oculomotor computation. Biol Cybern 97(4):279–292

104. Mueller SM, Paul WJ (2000) Computer architecture: complexity and correctness. Springer

105. Hennessy J, Patterson D (1996) Computer architecture: a quantitative approach, 2nd edn. Morgan Kaufmann Publishers

106. Le TQ, Truong TV, Tran HT, Park SH, Ko JH, Park HC, Byun D (2014) How could beetle's elytra support their own weight during forward flight? J Bionic Eng 11(4):529–540

107. Wang ZJ, Russell D (2007) Effect of forewing and hindwing interactions on aerodynamic forces and power in hovering dragonfly flight. Phys Rev Lett 99:148101

108. Xie CM, Huang WX (2015) Vortex interactions between forewing and hindwing of dragonfly in hovering flight. Theor Appl Mech Lett 5(1):24–29

109. Autumn K, Sitti M, Liang YA, Peattie AM, Hansen WR, Sponberg S, Kenny TW, Fearing R, Israelachvili JN, Full RJ (2002) Evidence for van der waals adhesion in gecko setae. Proc National Acad Sci 99(19):12252–12256

110. Montúfar G, Ghazi-Zahedi K, Ay N (2015) A theory of cheap control in embodied systems. PLoS Comput Biol 11(9):e1004427

111. Matsushita K, Lungarella M, C P, Yokoi H (2005) Locomoting with less computation but more morphology. In: Proceedings of 20th international conferenceon robotics and automation, pp 2020–2025

112. Pfeifer R, Gómez G (2009) Morphological computation—connecting brain, body, and environment. Springer, Berlin, Heidelberg, pp 66–83

113. Hauser H (2014) Morphological computation and soft robotics, shanghai lectures. http://shanghailectures.org/sites/default/files/guestlectures_slides/FINAL_ShanghAI_lecture_2013.pdf

114. Turing AM (1936) On computable numbers, with an application to the Entscheidungsproblem. Proc London Math Soc 2(42):230–265

115. Hopcroft JE, Motwani R, Ullman JD (2006) Introduction to automata theory, languages, and computation, 3rd edn. Addison-Wesley Longman Publishing Co., Inc, Boston, MA, USA

116. Piccinini G (2017) Computation in physical systems. In: Zalta EN (ed) The Stanford Encyclopedia of philosophy, summer, 2017th edn. Metaphysics Research Lab, Stanford University

117. Searle JR (1992) The rediscovery of the mind. MIT Press, Cambridge, MA, USA

118. Putnam H (1988) Representation and reality. MIT Press

119. Horsman C, Stepney S, Wagner RC, Kendon V (2014) When does a physical system compute? Proc Royal Soc A: Math, Phys Eng Sci 470(2169)

120. Reil T, Massey C (2003) Facilitating controller evolution in morpho-functional machines—a bipedal case study. Springer, Tokyo, Japan, pp 81–98

121. Mochon S, McMahon TA (1980) Ballistic walking: an improved model. Math Biosci 52(3):241–260

122. Scheutz M (2001) Computational versus causal complexity. Minds Mach 11(4):543–566

123. Jaeger H (2002a) Adaptive nonlinear system identification with echo state networks. In: Thrun S, Obermayer K (eds) Advances in neural information processing systems 15. MIT Press, Cambridge, MA, pp 593–600

124. Fernando C, Sojakka S (2003) Pattern recognition in a bucket. Springer, Berlin, Heidelberg, pp 588–597

125. Hermans M, Burm M, Van Vaerenbergh T, Dambre J, Bienstman P (2015) Trainable hardware for dynamical computing using error backpropagation through physical media. Nature Commun 6:6729 EP

126. Larger L, Soriano MC, Brunner D, Appeltant L, Gutierrez JM, Pesquera L, Mirasso CR, Fischer I (2012) Photonic information processing beyond turing: an optoelectronic implementation of reservoir computing. Opt Express 20(3):3241–3249

127. Paquot Y, Duport F, Smerieri A, Dambre J, Schrauwen B, Haelterman M, Massar S (2012) Optoelectronic reservoir computing. Sci Reports 2:287 EP

128. Brunner D, Soriano MC, Mirasso CR, Fischer I (2013) Parallel photonic information processing at gigabyte per second data rates using transient states. Nature Commun 4:1364 EP

129. Vandoorne K, Dierckx W, Schrauwen B, Verstraeten D, Baets R, Bienstman P, Campenhout JV (2008) Toward optical signal processing using photonic reservoir computing. Opt Express 16(15):11182–11192

130. Vandoorne K, Mechet P, Van Vaerenbergh T, Fiers M, Morthier G, Verstraeten D, Schrauwen B, Dambre J, Bienstman P (2014) Experimental demonstration of reservoir computing on a silicon photonics chip. Nature Commun 5:3541 EP

131. Kraukauer (2017) David Krakauer - q2. https://vimeo.com/125533384

132. Harris S (2016) Complexity & stupidity – a conversation with david krakauer. https://www.samharris.org/podcast/item/complexity-stupidity

133. World Cube Association (2018) World records, fewest moves. https://www.worldcubeassociation.org/

134. SpeedSolvingcom (2018) Rubik's cube speedsolving methods. https://www.speedsolving.com/wiki/index.php/Category:3x3x3_speedsolving_methods

135. Wolpert DH, Kolchinsky A, Owen JA (2017) The minimal hidden computer needed to implement a visible computation. https://arxiv.org/abs/1708.08494

136. Clark DD, Sokoloff L (1999) Circulation and energy metabolism of the brain. In: Siegel GJ, Agranoff BW, Albers RW, Fisher SK, Uhler MD (eds) Basic neurochemistry: molecular, cellular and medical aspects, 6th edn, Lippincott-Raven, Philadelphia, chap 31

Chapter 2
Information Theory—A Primer

I just wondered how things were put together.

Claude Shannon

The previous chapter discussed in detail how the body contributes to intelligence, discussed the term *Morphological Computation*, and finally, presented a new definition, called *Morphological Intelligence*. The next chapter will present different ways to quantify *Morphological Intelligence*, which are mainly based on information theory as it was defined by Shannon [1]. Hence, this chapter gives an introduction on the main information-theoretic concepts (based on [2]).

The first question is, why did we choose information theory to formalise the quantifications? There are several answers to this question. The most important reason is that information theory allows us to formulate fundamental principles very elegantly and almost independently of the underlying system. Shannon's definition of entropy (see Definition 2.1) is applied to communication, physics, natural language processing, to name just a few. Other reasons are that we can formulate high-level relationships as well as low-level relationships with the same formalism. Random variables (see below) can refer to symbols as well as spiking patterns of neurones. Hence, we can measure the entropy of words in a text in the same way we measure the entropy of spikes emitted by neurones. As a final reason, we argue that information theory for discrete systems is very illustrative, and hence, very comprehensible. This means that we can derive concepts on simple toy systems and apply them to more complex systems (with some restrictions that are discussed below). This is a very important aspect with respect to this book, because we are interested in understanding Morphological Intelligence from a formal perspective, but are equally interested in applying the concepts to real data.

This chapter is organised in the following way. We will first give a short primer on estimating probabilities on data. This will then be followed by a discussion of the fundamental concepts of information theory on discrete systems, i.e., entropy,

© Springer Nature Switzerland AG 2019
K. Ghazi-Zahedi, *Morphological Intelligence*,
https://doi.org/10.1007/978-3-030-20621-5_2

relative entropy, mutual information, and conditional mutual information. The final
section will discuss the estimation of these measures on real data based on the work
by Kraskov et al. [3] and Frenzel and Pompe [4]. Readers familiar with [2–4] can
skip this chapter.

2.1 Estimating Probabilities

Probabilities are estimated for random variables, so the first step is to define what
we mean by the term *random variable*. According to Blitzstein and Hwang [5],
a random variable is a variable whose possible values are numerical outcomes of
a random phenomenon. As mentioned in the introduction to this chapter, we use
the term random variable more generally, which means that a random variable can
also take symbolic values. We will use a car as an example. The random variable
describing its current motion could be symbolic and take the values `left`, `right`,
`stop`, ... The random variable describing the number of passengers is a discrete
number between 0 and 4. The random variable describing the state of the batteries
of our electric car is a continuous variable between 0 and 1.

Throughout this work, we denote random variables with capital letters, the real-
isation, i.e., the observed instance with lower-case letters and the alphabet, i.e., the
set of values that the random variable can take, with calligraphic letters. Hence, the
random variable X can take values $x \in \mathcal{X}$.

We will first discuss probabilities on discrete alphabets, i.e., $|\mathcal{X}| \in \mathbb{N}$.

A probability mass function assigns a probability to every value of a random
variable. It is defined in the following way:

$$p(x) \geq 0 \tag{2.1}$$

$$\sum_{x \in \mathcal{X}} p(x) = 1, \tag{2.2}$$

which means that the probability for each value is larger or equal to zero and that
all probabilities must sum up to 1. Probability mass functions are denoted by $p_X(x)$,
$p(X = x)$, or $p(x)$ for short. We will use $p(x)$ to denote the probability that the random
variable X will take the value $x \in \mathcal{X}$. If applicable, we will also write x instead of
$x \in \mathcal{X}$, when we sum over probabilities, i.e., Eq. (2.2) will be written as $\sum_x p(x) = 1$
in the remainder of this book.

The probability of a single variable is also called the marginal distribution. The
probability for more than one random variable is called the joint distribution. The
joint distribution of two random variables X and Y is denoted by $p(x, y)$. We will
use two perfect dice with six faces as an example in the rest of this section. Perfect
dice means that the dice will show each face with the same probability, and hence,
$p(x) = p(y) = \frac{1}{6}$ for both die, where X is the face of the first dice and Y is the
face of the second dice. Because both die are independent, the joint distribution is

$p(x, y) = p(x)p(y)$, i.e., the probability of each combination of X and Y is $p(x, y) = 1/6 \cdot 1/6 = 1/36$. The marginal distribution can be obtained from a joint distribution by summation over variables, i.e.

$$p(x) = \sum_y p(x, y), \text{ and} \qquad p(y) = \sum_x p(x, y). \qquad (2.3)$$

Conditional probability is a measure of the probability of an event given that another event was observed. In this book, we denote the conditional probability of X given that Y was observed by $p(x|y)$. Imagine the two dice discussed above, but now they are magically linked. Both dice show either an odd or even number, but never mixed, i.e., if the first dice shows 1, 3, or 5, then the second die will also show an odd number. The conditional probability for our two die is then given by $p(X = i|Y = j) = 1/3$ if $i, j \in \{1, 3, 5\}$ or $i, j \in \{2, 4, 6\}$ and it is zero otherwise. Note that for each condition, the probabilities must sum up to one, i.e.,

$$\forall y: \sum_x p(x|y) = 1. \qquad (2.4)$$

Please also note, that a conditional probability does not imply causation. This difference between conditional probability and the causal dependence between two random variables will be discussed below (see Sect. 2.7). The final remark with respect to probabilities is the relation between conditional and joint distributions, which is given by

$$p(x, y) = p(x|y)p(y). \qquad (2.5)$$

The reader interested in a comprehensive introduction to probability theory is referred to e.g. [6]. The question now is, how can these marginal, joint, and conditional probabilities be estimated from data. The underlying problem is visualised in Fig. 2.1.

The problem is explained based on the example of human height. We don't know the underlying probability density function of human height, i.e.m, we don't know the average height of humans and the distributions of heights overall. We call the underlying probability density distribution $\mu(x)$, where x is the height of a person. To estimate $\mu(x)$, we have to select a group of people and measure their height. From this data (samples), we then have to estimate $\mu(x)$.

The problem is formulated in the following way: We have to estimate a continuous probability density function $\mu(x)$ based on limited samples (see Fig. 2.1a). The most commonly used method is known as *binning* (see Fig. 2.1b). Here, the state space is discretized into bins, where each bin is defined by its domain. Usually, bins are equally distributed in the state space, which means that they all have the same size. This estimation method is based on counting how many samples x_t fall into each the i-th bin, where t refers to time. We denote the number of samples in bin i with c_i. The probability mass function $p(x)$ is then given by:

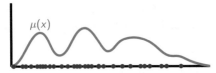

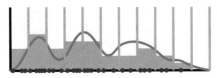

(a) Goal: Estimating a continuous probability density function $\mu(x)$ based on limited samples.

(b) The most commonly used approach is known as *Binning* or *Frequency-based estimation*.

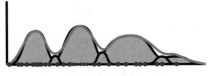

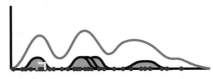

(c) Another very popular method is known as *Mixture of Gaussians*.

(d) Estimation with k-nearest neighbour. The distance to the k-nearest neighbour is used to parametrise, e.g. a constant, or as depicted here, a Gaussian kernel.

Fig. 2.1 Estimating Probabilities. **a** Goal: Estimating continuous probabilities density function $\mu(x)$ based on limited samples. **b** *Frequency-based method*: In this approach, the state space is discretised, usually into bins of equal size. For each bin, the number of samples that fall into that bin are counted and then normalised by the total number of samples. The resulting number is the estimated probability. The quality of the estimation depends on the bin size and the number of samples. **c** *Mixture of Gaussians*: This method uses a, usually pre-defined, number of Gaussian kernels to estimate $\mu(x)$. **d** k-nearest neighbour method is similar to the *Mixture of Gaussian kernels* method in that it uses mixtures of either constant functions or Gaussian kernels. The width of each function is determined by the distance to the k-nearest neighbour. This method is used in this book (see Sect. 2.8)

$$p(x_i) = \frac{c_i}{N}, \tag{2.6}$$

where N is the total number of samples. The law of large numbers states that with enough samples, $p(x)$ will converge towards $\mu(x)$, i.e., $\lim_{N \to \infty} p(|p(x_i) - \mu(x_i)| > \varepsilon) = 0$ [7]. This estimator is also known as the empirical estimator. The problem with this method is, that the number of samples is considerably small in most applications, especially if the number of samples is compared with the dimensionality of the state space.

The second method that is discussed here is designed to overcome this problem by mixing Gaussian kernels (see e.g. [8, 9]). Figure 2.1c) illustrates the underlying idea. A number of Gaussian kernels are distributed over the state space such that their mixture (e.g. sum) approximates $\mu(x)$. The estimated probability distribution $\tilde{\mu}(x)$ is then given by a mixture of all Gaussian kernels:

$$\tilde{\mu}(x) = \sum_i \phi_i g_i(x, b_i, \sigma_i), \tag{2.7}$$

where $g_i(x, b_i, \sigma_i) = (\sigma_i \sqrt{2\pi}) e^{-\frac{1}{2}\left(\frac{x-b_i}{\sigma}\right)^2}$ is a Gaussian and ϕ_i is a free parameter. The advantage of this method is that it is a continuous estimation of probability distributions. The disadvantage is that the number of kernels is an open parameter that is usually pre-determined.

The final method uses the distance to the k-nearest neighbour as an approximation for local probability densities. Instead of pre-defining the number of kernels as in the previous approach, a Gaussian Kernel (or constant function) is placed at the location of each sample (each sampled point in the state space). The parameters of the local approximation function (e.g. Gaussian or constant function) are then approximated based on the distance to the k-nearest neighbour. The advantage of this approach is that it can be used directly to estimate entropy, mutual information, etc. without an additional discretisation step. This will be explained in more detail below (see Sect. 2.8).

2.2 Summary

In this section, we have discussed a few basic concepts of probability theory. The first concept was a random variable, which is a variable that captures the outcome of some stochastic phenomenon. In the context of this book, random variables can take numerical values and symbols as their value. We then discussed marginal, joint, and conditional probabilities, how they relate to each other and closed this section with a brief overview of different methods how probabilities can be estimated from samples.

The next sections will present fundamental concepts of information theory, first for discrete state spaces and then for continuous state spaces.

2.3 Entropy

The most fundamental concept in information theory [1] is entropy. Entropy is a measure of the uncertainty of a random variable [2]. The entropy of a random variable X is defined in the following way.

Definition 2.1 (*Entropy*) The entropy of a discrete random variable X is defined as

$$H(X) = -\sum_x p(x) \log p(x) \qquad (2.8)$$

In the remainder of this book, we will only use the logarithm with base 2, i.e., all measures will be given with $\log_2$, which means that the results of the measure are given in bits. The basis can be changed easily, as the following calculation shows:

$$\log_b p = \log_b a \log_a p \tag{2.9}$$

$$\Rightarrow H_b(X) = -\sum_x p(x)(\log_b a) \log_a p(x) \tag{2.10}$$

$$= (\log_b a)\left(-\sum_x p(x) \log_a p(x)\right) \tag{2.11}$$

$$= (\log_b a) H_a(X) \tag{2.12}$$

An important observation is that the entropy of a random variable is always non-negative, i.e. $H(X) \geq 0$. This follows from the fact that probabilities are non-negative and smaller or equal to one, i.e., $0 \leq p(x) \leq 1$ which leads to $\log p(x) \leq 0$, and hence, $H(X) = -\sum p(x) \log p(x) \geq 0$.

Entropy can also be viewed as the expected surprise of a random variable X. If $p(x)$ is the probability of an even $x \in \mathcal{X}$, then $-\log_2 p(x)$ can be defined as the surprise of this event, i.e., the surprise is maximal if the likelihood is minimal and vice versa. This leads to the definition of entropy as the expected surprise:

$$H(X) = -\sum_x p(x) \log_2 p(x) \tag{2.13}$$

$$= \mathbb{E}[-\log_2 p(x)]. \tag{2.14}$$

The joint entropy of two random variables is defined in the following way:

Definition 2.2 (*Joint Entropy*) The joint entropy of a pair of random variables X, Y with joint distribution $p(x, y)$ is defined as

$$H(X, Y) = -\sum_{x,y} p(x, y) \log p(x, y) \tag{2.15}$$

Finally, the conditional entropy of a random variable X given another random variable Y is defined as:

Definition 2.3 (*Conditional Entropy*) Given two discrete random variables X, $Y \sim p(x, y)$, then the conditional entropy is defined as

$$H(X|Y) = \sum_y p(y) H(X|Y = y) \tag{2.16}$$

$$= -\sum_y p(y) \sum_x p(x|y) \log p(x|y) \tag{2.17}$$

$$= -\sum_{x,y} p(x, y) \log p(x|y) \tag{2.18}$$

The three quantities entropy, conditional entropy, and joint entropy are related inductively. The joint entropy of two variables is the sum of the uncertainty of one of the variables and the uncertainty of the second given the first, i.e.,

$$H(X, Y) = H(X) + H(Y|X). \tag{2.19}$$

This is also known as the *Chain Rule for Bivariate Entropy* [2]. The following calculations show that the chain rule is valid for the definitions given above:

$$H(X, Y) = -\sum_x \sum_y p(x, y) \log p(x, y) \tag{2.20}$$

$$= -\sum_x \sum_y p(x, y) \log(p(y|x)p(x)) \tag{2.21}$$

$$= -\sum_x \sum_y p(x, y) \log p(x) - \sum_x \sum_y p(x, y) \log p(y|x) \tag{2.22}$$

$$= -\sum_x p(x) \log p(x) - \sum_x \sum_y p(x, y) \log p(y|x) \tag{2.23}$$

$$= H(X) + H(Y|X). \tag{2.24}$$

This confirms that the definitions of entropy, conditional entropy and joint entropy are natural [2].

The generalisation of the chain rule for bivariate entropy is known as the *Chain Rule for Multivariate Entropy* and it is given by the following equation:

$$H(X_1, X_2, \ldots, X_n) = \sum_{i=1}^{n} H(X_i|X_{i-1}, \ldots, X_1) \tag{2.25}$$

The chain rule for multivariate entropy can be proven by applying the chain rule for bivariate entropy (see Eq. (2.19)) repeatedly, as the following calculations show:

$$H(X_1, X_2) = H(X_1) + H(X_2|X_1) \tag{2.26}$$
$$H(X_1, X_2, X_3) = H(X_1) + H(X_2, X_3|X_1) \tag{2.27}$$
$$H(X_1, X_2, X_3) = H(X_1) + H(X_2|X_1) + H(X_3|X_2, X_1) \tag{2.28}$$

$$\vdots \tag{2.29}$$

$$H(X_1, X_2, \ldots, X_n) = H(X_1) + H(X_2|X_1) + \cdots + H(X_n|X_{n-1}, \ldots, X_1) \tag{2.30}$$

$$= \sum_{i=1}^{n} H(X_i|X_{i-1}, \ldots, X_1) \tag{2.31}$$

One of the most important questions in the context of this book is how well an observation is explained by a model. We will call the observation p and the model q. Given the example of our two dice above, such a question could be how well the statistics of the faces is explained by the assumption that the two die are fair and independent. In this case, our model q is the bivariate model $q(x, y) = p(x)p(y)$, where $p(x) = p(y) = 1/6$. Our observation is the joint distribution $p(x, y)$ which could have been estimated e.g. with the empirical method (see Eq. (2.6)). We now want to measure the difference of these two distributions. This is also known as the *Kullback-Leibler Divergence* or *Relative Entropy*. It is defined in the following way:

Definition 2.4 (*Kullback-Leibler Divergence*) The *Kullback-Leibler Divergence* (KL-distance, KL-divergence or relative entropy) between two probability mass functions $p(x)$ and $q(x)$ is defined as

$$D_{KL}(p||q) = \sum_x p(x) \log \frac{p(x)}{q(x)} \qquad (2.32)$$

The following conventions are used $0 \log \frac{0}{0} = 0$, $0 \log \frac{0}{q} = 0$, and $p \log \frac{p}{0} = \infty$. This means that if there is a symbol $x \in \mathcal{X}$ for which $p(x) > 0$ and $q(x) = 0$, then $D(p||q) = \infty$.

An important property of the Kullback-Leibler distance is that it is always nonnegative. The prove requires Jensen's inequality which makes a statement about the expectation of *convex functions*. A function $f(x)$ is said to be convex over an interval (a, b) if for every $x_1, x_2 \in (a, b)$ and $0 \leq \lambda \leq 1$ the following inequality is satisfied

$$f(\lambda x_1 + (1 - \lambda)x_2) \leq \lambda f(x_1) + (1 - \lambda)f(x_2). \qquad (2.33)$$

A function f is said to be *strictly convex* if equality holds only for $\lambda = 0$ or $\lambda = 1$.

Let $\mathbb{E}$ denote the *expectation*, i.e., $\mathbb{E}X = \sum_x p(x) x$ in the discrete case and $\mathbb{E}X = \int xf(x)dx$ in the continuous case. We can now state Jensen's inequality [2] in the following way:

Theorem 2.1 (Jensen's Inequality) *If f is a convex function and X is a real-valued random variable, then*

$$\mathbb{E}f(X) \geq f(\mathbb{E}X). \qquad (2.34)$$

Moreover, if f is strictly convex, the equality in Eq. (2.34) implies that $X = \mathbb{E}X$ with probability 1, i.e., X is a constant.

We prove this inductively by proving the theorem for two points and inducing over the number of points. This implies that this proof is given for discrete systems only.

Equation (2.34) for two points is

$$p_1 f(x_1) + p_2 f(x_2) \geq f(p_1 x_1 + p_2 x_2). \qquad (2.35)$$

This follows directly from the definition of convex functions (see Eq. (2.33)). We now assume that Jensen's inequality is true for $k - 1$ points, and define $p_i' = p_i/(1 - p_k)$ for $i = 1, 2, \ldots, k - 1$. It follows that

$$\sum_{i=1}^{k} p_i f(x_i) = p_k f(x_k) + (1 - p_k) \sum_{i=1}^{k-1} p_i' f(x_i) \tag{2.36}$$

$$\geq p_k f(x_k) + (1 - p_k) f\left(\sum_{i=1}^{k-1} p_i' x_i\right) \tag{2.37}$$

$$\geq f\left(p_k x_k + (1 - p_k) \sum_{i=1}^{k-1} p_i' x_i\right) \tag{2.38}$$

$$= f\left(\sum_{i=1}^{k} p_i x_i\right). \tag{2.39}$$

The first inequality (Eq. (2.37)) follows from the induction hypothesis and the second inequality (Eq. (2.38)) follows from the definition of convexity (see Eq. 2.33).

We can now use Jensen's inequality to prove the non-negativity of the Kullback-Leibler distance. Let $p(x), q(x), x \in \mathcal{X}$ be two probability mass functions. Then

$$D(p\|q) \geq 0 \tag{2.40}$$

with equality if and only if $p(x) = q(x)$ for all x. This is also known as *Information Inequality*.

The proof requires a division by $p(x)$, and hence, we have to restrict ourselves those values of $p(x)$ which are strictly positive. This set is also known as the support of $p(x)$ and it is defined in the following way: $A = \{x : p(x) > 0\}$. We can now prove the information inequality on the support using Jensen's inequality:

$$-D(p\|q) = -\sum_{x \in A} p(x) \log \frac{p(x)}{q(x)} \tag{2.41}$$

$$= \sum_{x \in A} p(x) \log \frac{q(x)}{p(x)} \leq \log \sum_{x \in A} p(x) \frac{q(x)}{p(x)} \tag{2.42}$$

$$= \log \sum_{x \in A} q(x) \leq \log \sum_{x \in \mathcal{X}} q(x) \tag{2.43}$$

$$= \log 1 \tag{2.44}$$

$$= 0. \tag{2.45}$$

The first inequality (see Eq. (2.42)) follows from Jensen's inequality and the fact that log is a strictly convex function (strictly positive second derivative, which is not further discussed here). The Kullback-Leibler divergence is zero ($D(p\|q) = 0$) if Eqs. (2.42) and (2.43) are equalities. To understand when the KL-Divergence is zero, we need to understand the cases in which the two equations are equalities. Equation (2.42) is an equality if $q(x)/p(x)$ is a constant everywhere, i.e., $q(x) = c\,p(x)$ for all x. Hence, $\sum_{x \in A} q(x) = c \sum_{x \in A} p(x) = c$. It then follows that Eq. (2.43) is an equality if $\sum_{x \in A} q(x) = \sum_{x \in \mathcal{X}} q(x) = c$, which implies that $c = 1$. This means that we only have equality in both equations only if the two distributions are the same, i.e., $p(x) = q(x)$.

In the context of this book, we will often use the *conditional relative entropy*, which is the Kullback-Leibler Divergence for conditional distributions. It is defined in the following way.

Definition 2.5 (*Conditional Relative Entropy*) For joint probability mass functions $p(x, y)$ and $q(x, y)$, the conditional relative entropy $D(p(y|x)||q(y|x))$ is the average of the relative entropies between the conditional probability mass functions $p(y|x)$ and $q(y|x)$ averaged over the probability mass function $p(x)$. More precisely,

$$D(p(x|y)||q(x|y)) = \sum_x p(x) \sum_y p(y|x) \log \frac{p(x|y)}{q(x|y)}. \tag{2.46}$$

Non-negativity of the conditional relative entropy follows from the information inequality (see Eqs. (2.40)) and (2.5), i.e., $D(p(x|y)||q(x|y)) \geq 0$. Analogous to the chain rule for entropies (see Eq. (2.25)), we can also derive the chain rule for relative entropies in the following way:

$$D(p(x, y)||q(x, y)) = \sum_x \sum_y p(x, y) \log \frac{p(x, y)}{q(x, y)} \tag{2.47}$$

$$= \sum_x \sum_y p(x, y) \log \frac{p(y|x)p(x)}{q(y|x)q(x)} \tag{2.48}$$

$$= \sum_x \sum_y p(x, y) \log \frac{p(x)}{q(x)} \tag{2.49}$$

$$+ \sum_x \sum_y p(x, y) \log \frac{p(y|x)}{q(y|x)} \tag{2.50}$$

$$= D(p(x)||q(x)) + D(p(y|x)||q(y|x)) \tag{2.51}$$

So far, we have defined entropy, joint entropy, conditional entropy, relative entropy, and finally, conditional relative entropy. These quantifications are the foundation for all measures defined in this and the following chapters.

It is important to know maximal values that these quantifications can achieve in some cases, e.g. normalisation of measures for comparison. We can derive the upper bound for the entropy with Jensen's inequality in the following way.

Let $u(x) = 1/|\mathcal{X}|$ be the uniform probability mass function over $\mathcal{X}$, and let $p(x)$ be the probability mass function for X. Then (using Jensen's inequality once again):

$$D(p||u) = \sum_x p(x) \log \frac{p(x)}{u(x)} \tag{2.52}$$

$$= \sum_x p(x) \log p(x) - \sum_x p(x) \log u(x) \tag{2.53}$$

$$= \sum_x p(x) \log p(x) - \sum_x p(x) \log \frac{1}{|\mathcal{X}|} \tag{2.54}$$

$$= \sum_x p(x) \log p(x) - \log \frac{1}{|\mathcal{X}|} \tag{2.55}$$

$$= \log |\mathcal{X}| - H(X) \tag{2.56}$$

Hence, because of the information inequality (non-negativity of the relative entropy, see Eq. (2.40)), it follows that

$$0 \le D(p||u) = \log |\mathcal{X}| - H(X), \tag{2.57}$$

with equality if and only if $p = u$, i.e., X is uniformly distributed. This means that the entropy $H(X)$ is upper bounded by

$$H(X) \le \log |X|. \tag{2.58}$$

Summary

Entropy is a measure of uncertainty of a random variable, which can also be understood as the expected surprise. The conditional entropy describes the uncertainty of one random variable in the context that the outcome of a second variable is known. It is the expected surprise of the conditional probability. The joint entropy of two random variables is given by the sum of the marginal and conditional entropies, which can be generalised to the chain rule. The basis for most quantifications in the remainder of this book is the relative entropy, which can be understood as the difference between an observation p and a model q. It is also known as Kullback-Leibler Divergence or KL-Divergence. Entropies are upper bounded by the logarithm of the alphabet's cardinality and always non-negative.

This concludes the section on entropies. The following sections will discuss mutual information and conditional mutual information, after giving a brief summary.

2.4 Mutual Information

Mutual information is the amount of information that one variable X contains about another variable Y. In the example of our two magically connected dice above, knowledge about one die contains information about the other die, because they both either show an odd or even number. If I know the face of one die, I know that the other die can only have one of three faces. This means that my uncertainty about the second die is reduced due to information that was provided by the first die. Shannon [1] was interested in modelling communication over noisy channels. In this context, the random variable X was the message transmitted by a sender and the random variable Y is the message received by the recipient over a noisy communication channel. One of the big achievements of Shannon was to create the mathematical framework that allowed to maximise the mutual information of the received message Y and the emitted message X over a noisy communication channel.

Revisiting the example above, we would say that there is no mutual information for two regular, uncoupled dice, i.e. for the two dice that show each face with equal probability We would argue that maximal mutual information is given by the case in which one die always is a copy of the other die, because in this case, knowing one die removes all uncertainty about the other.

In the context described above (see relative entropy, Definition 2.4) mutual information can be understood as the question of how well the observation of two random variables can be explained by the assumption that the random variables are independent. This leads to the following definition.

Definition 2.6 (*Mutual Information*) Let X and Y denote two random variables with a joint probability mass function $p(x, y)$ and marginal probability mass functions $p(x)$ and $p(y)$. The mutual information $I(X; Y)$ is the relative entropy between the joint distribution $p(x, y)$ and the product distribution $p(x)p(y)$:

$$I(X; Y) = D_{KL}(p(x, y) \| p(x)p(y)) \tag{2.59}$$

$$= \sum_{x,y} p(x, y) \log \frac{p(x, y)}{p(x)p(y)} \tag{2.60}$$

The mutual information $I(X; Y)$ is always non-negative and zero only if X and Y are independent, which is often denoted by $X \perp\!\!\!\perp Y$. The information inequality (see Eq. (2.40)) states that $D(p\|q) \geq 0$, with equality if and only if $p = q$. Hence, it follows that $I(X; Y) = D(p(x, y)\|p(x)p(y)) \geq 0$ with equality if and only if $p(x, y) = p(x)p(y)$, i.e., if X and Y are independent.

In the following paragraph, we will discuss a few important properties of mutual information. The first property is used frequently in this book. It states that the mutual information of two random variables can be expressed in terms of entropy and conditional entropy in the following ways:

$$I(X; Y) = H(X) - H(X|Y) \tag{2.61}$$
$$I(X; Y) = H(Y) - H(Y|X) \tag{2.62}$$

To prove this, we use Eq. (2.60):

$$I(X;Y) = \sum_{x,y} p(x,y) \log \frac{p(x,y)}{p(x)p(y)} \tag{2.63}$$

$$= \sum_{x,y} p(x,y) \log \frac{p(x|y)}{p(x)} \tag{2.64}$$

$$= \sum_{x,y} p(x,y) \log p(x|y) - \sum_{x,y} p(x,y) \log p(x) \tag{2.65}$$

$$= H(X) - H(X|Y) \tag{2.66}$$

Equation (2.62) follows from Eq. (2.64) by symmetry:

$$\frac{p(x,y)}{p(x)p(y)} = \frac{p(x|y)}{p(y)} = \frac{p(y|x)}{p(x)}. \tag{2.67}$$

Mutual information can also be expressed as a sum of marginal and joint entropies in the following way:

$$I(X;Y) = H(X) + H(Y) - H(X,Y). \tag{2.68}$$

Equation (2.68) follows from Eq. (2.66) and the chain rule (see Eq. (2.24)).

The final two properties are symmetry

$$I(X;Y) = I(Y;X) \tag{2.69}$$

and the fact that the mutual information of a variable with itself is the entropy of the variable:

$$I(X;X) = H(X). \tag{2.70}$$

Symmetry (see Eq. (2.69)) follows from Eq. (2.67) and equality with entropy for a single variable (see Eq. (2.70)) follows from Eq. (2.61):

$$I(X;X) = H(X) - H(X|X) = H(X). \tag{2.71}$$

From Eq. (2.61) we know that $I(X;Y) = H(X) - H(X|Y)$ and from Eq. (2.40) we know that $I(X;Y) > 0$, hence, it follows that conditioning reduces entropy, since:

$$0 \le I(X;Y) = H(X) - H(X|Y) \tag{2.72}$$
$$\rightarrow H(X|Y) \le H(X), \tag{2.73}$$

with equality in Eq. (2.73) if and only if X and Y are independent, i.e., $I(X;Y) = 0$.

Summary

This concludes the section on mutual information. To summarise, mutual information is the amount of information that one random variable contains about another. It is equal to zero only if the two random variables are independent and positive otherwise. It is upper bounded by the univariate (single-value) entropy of both random variables and it is symmetric. Finally, it can be expressed in terms of conditional, joint, and marginal entropies.

This concludes the section on mutual information.

2.5 Conditional Mutual Information

Conditional mutual information is the information that one variable contains about another given that the outcome of a third variable is known. Conditional mutual information is used frequently in the next chapter. It is defined in the following way:

Definition 2.7 (*Conditional Mutual Information*) The *conditional mutual information* of random variables X and Y given Z is defined by

$$I(X; Y|Z) = H(X|Z) - H(X|Y, Z) \tag{2.74}$$

$$= \sum_{x,y,z} p(x, y, z) \log \frac{p(x, y|z)}{p(x|z)p(y|z)} \tag{2.75}$$

The *chain rule for information* shows how multivariate mutual information can be decomposed into a chain of conditional mutual informations:

$$I(X_1, X_2, \ldots, X_n; Y) = H(X_1, X_2, \ldots, X_n) - H(X_1, X_2, \ldots, X_n|Y) \tag{2.76}$$

$$= \sum_{i=1}^{n} H(X_i|X_{i-1}, \ldots, X_1) - \sum_{i=1}^{n} H(X_i|X_{i-1}, \ldots, X_1, Y) \tag{2.77}$$

$$= \sum_{i=1}^{n} I(X_i; Y|X_{i-1}, \ldots, X_1) \tag{2.78}$$

Conditional mutual information is non-negative (follows from the Information Inequality, see Eq. (2.40)) and upper bounded by the entropy, i.e.,

$$I(X; Y|Z) \geq 0. \tag{2.79}$$
$$I(X; Y|Z) = H(X|Y) - H(X|Y, Z) \tag{2.80}$$
$$\leq H(X|Y) \tag{2.81}$$
$$\leq H(X) \tag{2.82}$$
$$\leq \log |X|. \tag{2.83}$$

Inequality (2.82) follows from Eq. (2.73).

We will close this section with an overview of relationships between conditional mutual information, mutual information and entropy:

$$I(X;Y|Z) = H(X,Z) + H(Y,Z) - H(X,Y,Z) - H(Z) \qquad (2.84)$$

$$I(X;Y|Z) = H(X|Z) - H(X|Y,Z) \qquad (2.85)$$

$$I(X;Y|Z) = H(X|Z) + H(Y|Z) - H(X,Y|Z) \qquad (2.86)$$

$$I(X;Y|Z) = I(X;Y,Z) - I(X;Z) \qquad (2.87)$$

This concludes the section about conditional mutual information.

This book is concerned with the comparison of control complexity and behavioural complexity. A canonical candidate in this context is the Kolmogorov complexity, which is discussed next.

2.6 Kolmogorov Complexity

Kolmogorov complexity is the question of finding the algorithmic complexity of an object, which is the shortest binary computer program that describes the object (usually a binary string). In the next chapter, we will talk about the complexity of behaviour and the complexity of a control program (policy), which is strongly related to Kolmogorov complexity. Hence, it is important to discuss Kolmogorov Complexity and show why it is not applied in this book. For this purpose, we will introduce Kolmogorov complexity as far as necessary and present a proof that it is not computable. This fact is the reason why Kolmogorov complexity is not considered in this book.

Let x be a finite-length binary string and let U be a universal computer (e.g. Turing's universal computer, see Sect. 1.2.3). Let $l(x)$ denote the length of the string x. Let $U(p)$ denote the output of the computer U when presented with a program p.

Kolmogorov complexity of the string x is then defined as the minimal description length of x.

Definition 2.8 (*Kolmogorov Complexity*) The Kolmogorov complexity $K_U(x)$ of a string x with respect to a universal computer U is defined as

$$K_U(x) = \min_{p:U(p)=x} l(p), \qquad (2.88)$$

which is the minimum length over all programs p that print x and halt. Thus, $K_U(x)$ is the shortest description length of x over all descriptions interpreted by computer U that result in x.

A very important property of Kolmogorov complexity is that it is independent of the computer it runs on or the language that is used to describe the program. This is also known as the *Universality of Kolmogorov Complexity* At first, this seems counter-intuitive because one can easily imagine programs that written in less code depending on the utilised programming language. To understand why Kolmogorov Complexity is independent of the computer, imagine a sort of emulator that runs on computer A (or in programming language A). An emulator is a program that runs on one computer and simulates the architecture of another computer. The Kolmogorov Complexity of the emulator that runs on computer A and is able to emulate computer B is independent of the actual program that we want to emulate since it is able to emulate all programs of computer B on computer A. We can now estimate the complexity of a program written for computer (language) B that runs on computer B in the following way.

We denote the program written for computer A to print string x with p_A, i.e., $A(p_A) = x$. The emulation of computer A written for computer U is denoted by s_U^A. The total length of the program written for U to print x is then given by:

$$l(p) = l(s_U^A) + l(p_A) = c_A + l(p_A), \tag{2.89}$$

where c_A is the length of the required emulation. It then follows that

$$K_U(X) = \min_{p:U(p)=x} l(p) \leq \min_{p:A(p)=x}(l(p) + c_A) = K_A(x) + c_A \tag{2.90}$$

for all binary strings x. Hence, we can make the following statement:

$$K_U(x) \geq K_A(x) + c_A \tag{2.91}$$

for all binary strings and the constant c_A does not depend on x, which is also known as the *Universality of Kolmogorov Complexity*.

It should be obvious now, why Kolmogorov Complexity is a very appealing concept, especially in the context of this book. The reason is that are interested in studying and comparing the complexity of behaviours and policies, i.e., descriptions of how an agent has behaved in the environment and the control programs that generated this behaviour. With Kolmogorov Complexity, one could ask the question how complex it would be to in-source (see Sect. 3.6) complexity from the morphology into the brain or outsource computation from the brain into the physics of the body and body-environment interactions. Unfortunately, it can be easily proven that Kolmogorov complexity is not computable, which means that it cannot be applied.

To understand what this means, assume that we have a program that can compute the Kolmogorov Complexity of a string x, named `kolmogorov(x)` that has Kolmogorov Complexity $K(\texttt{kolmogorov(x)}) = m$. We can now write a program that generates strings with Kolmogorov complexity $m + n$ that is less complex than $m + n$ in the following way (see Algorithm 2.1).

This program prints every string that is more complex (with respect to the Kolmogorov complexity) than the description of this program requires. This is a contradiction and shows that Kolmogorov complexity is not computable, because we can find a program that is shorter than the Kolmogorov complexity of its output would determine. This means that for a given string, we cannot compute its Kolmogorov complexity. Applied to the context of this work, we cannot compute the Kolmogorov

Algorithm 2.1 Proof that $K(x)$ is not computable. This program returns all strings that have a Kolmogorov complexity $K(x) \geq n + m$, although the Kolmogorov complexity of the program is $n + m$. This is a contradiction and shows that the Kolmogorov complexity is not computable.

Require: $m \leftarrow$ Kolmogorov complexity of this program
Require: $n \leftarrow$ Kolmogorov complexity of the function `kolmogorov(x)`
  for $i = 1$ to ∞ **do**
    for each string x of length $l(x) = i$ **do**
      if `kolmogorov(x)` $> m + n$ **then**
        print x
      end if
    end for
  end for

complexity of the control program or behaviour, which is why Kolmogorov Complexity will not be considered in the next chapter.

2.7 Causality Versus Correlation

Conditional probabilities (see above) are often treated as if they capture the causal dependency between two random variables. This section discusses the difference between causation and correlation in the context of information and probability theory based on Pearl's causality theory [10].

Consider the following system consisting of four random variables A, B, C, and D, where B is causally dependent on A, C is causally dependent on B, and finally D is causally dependent on A and C. By causally dependent we meant that there is a mechanism that connects the two random variables, such that if e.g. A changes its value, this change has a direct effect on B. The system is fully described by the probability distribution $p(a)$ and the mechanism $\gamma(b|a)$, $\eta(c|b)$, and $\phi(d|a, b)$. The joint distribution of the entire system is given by $p(a, b, c, d) = p(a)\, \gamma(b|a)\, \eta(c|b)\, \phi(d|a, b)$. We can now calculate the conditional probability distribution (for $p(d) > 0$)

$$p(d|b) = \sum_{a,c} \frac{p(a, b, c, d)}{p(d)} \tag{2.92}$$

$$= \sum_{a,c} \frac{p(a)\, \gamma(b|a)\, \eta(c|b)\, \phi(d|b, a)}{p(d)}, \tag{2.93}$$

which might be larger than zero for most values of b and d.

Let us now assume that the causal link between B and C is removed (shown in Fig. 2.2, right-hand side). In this case, the conditional probability distribution $p(d|b)$ can still be computed in the following way

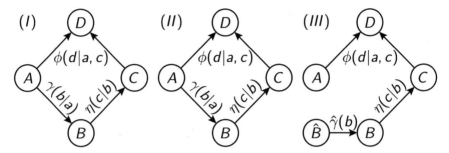

Fig. 2.2 *Virtual Intervention.* This figure shows the concept of virtual intervention as it is defined by Pearl [10]. To investigate if the random variable B has a causal effect of D, the causal link between A and B is replaced by a causal link between a new random variable $\hat{B}$ and B. The new random variable $\hat{B}$ refers to the virtual intervention. This is explained in detail in the text below

$$p(d\,|b) = \sum_{a,c} \frac{p(a,b,c,d)}{p(d)} \tag{2.94}$$

$$= \sum_{a,c} \frac{p(a)\,\gamma(b)\,\tilde{\eta}(c|b)\,\phi(d|b,a)}{p(d)} \tag{2.95}$$

and, because of the common cause A, will likely lead to positive, non-zero values. The reason is that $p(d|b)$ is a measure of correlation that is not able to distinguish between e.g. a common cause and causal dependence.

One way to distinguish between a common cause and a direct causal dependence is experimental intervention. Experimental intervention means that an experimenter actively changes B and observes how this affects D. An example is a medical test in which one group is handed the medicine that is tested and another group is given a placebo. This is an active intervention in B to investigate different outcomes in D. The question now is, can we post hoc determine if the jersey actually had a causal influence on the outcome of the game, given that we have enough data points? This is called post hoc or virtual intervention and it asks the question, what would have happened in D if I had changed B.

The formalism that we use in this work is based on Pearl's causality theory [10] that was also investigated by Ay and Polani [11] and discussed in the context of the sensorimotor loop by [12]. Hence, in the following paragraphs, we will only describe it as far as it is necessary to understand the remainder of this section. For full mathematical treatment, the reader is referred to [10–12].

Intervening post hoc in B corresponds to virtually adding a new random variable $\hat{B}$, which is virtually connected to the random variable B, thereby replacing the connection from A (see Fig. 2.2c). This means that we have replaced the mechanism $\gamma(b|a)$ with the mechanism $\hat{\gamma}(b)$ that directly sets the values of B. The new mechanism is defined such that it is equal to one if b has the desired value $\hat{b}$ (the intervention) and zero otherwise:

$$\hat{\gamma}(b) = \delta_{b\hat{b}} = \begin{cases} 1, & b = \hat{b} \\ 0, & b \neq \hat{b} \end{cases}. \tag{2.96}$$

The post-interventional joint distribution, denoted by $p_{\hat{b}}(a, b, c, d)$ is now defined as:

$$p_{\hat{b}}(a, b, c, d) = p(a)\,\hat{\gamma}(b)\,\tilde{\eta}(c|b)\phi(d|b, a) \tag{2.97}$$

We can then define the *do* operator for intervention according to Pearl [10] in the following way:

$$p(a, c, d|\mathrm{do}(b)) := \sum_{b'} p_{\hat{b}}(a, b', c, d) \tag{2.98}$$

$$= \sum_{b'} p(a)\,\hat{\gamma}(b')\,\tilde{\eta}(c|b')\,\phi(d|b, a) \tag{2.99}$$

Together with Eq. (2.96), this results in

$$p(a, c, d|\mathrm{do}(b)) = p(a)\,\tilde{\eta}(c|b')\,\phi(d|b, a) \tag{2.100}$$

In a nutshell, this means that a post hoc intervention is equivalent to removing the conditional probability distributions from the model that conditions on the random variable which is the target of the intervention. This is only possible if the parents of the removed random variable are known (in this case A).

This formalism can now be used to determine the causal information flow from one random variable X to another random variable Y, (in this book denoted by $CIF(X \rightarrow Y)$, which is different than e.g. calculating the mutual information of both variables. The conditional information flow is the Kullback-Leibler divergence of the post-interventional probability distribution $\hat{p}(y)$ and the post-interventional conditional probability distribution $p(y|\hat{\mathrm{do}}(x))$:

$$CIF(X \rightarrow Y) = \sum_{x,y} \hat{p}(y|\mathrm{do}(x))p(x)\log_2 \frac{\hat{p}(y|\mathrm{do}(x))}{\hat{p}(y)} \tag{2.101}$$

$$= \sum_{x,y} \hat{p}(y|\mathrm{do}(x))p(x)\log_2 \frac{\hat{p}(y|\mathrm{do}(x))}{\sum_x \hat{p}(y|x)p(x)}. \tag{2.102}$$

For a detailed discussion on this topic, please read [11].

This closes the discussion on information theory on discrete state spaces. The next section will cover estimators for entropy on continuous state spaces.

2.8   Entropy Estimation on Continuous State Spaces

In this book, we are interested in quantifying Morphological Intelligence in the context of embodied systems. These can either be robots or biological systems, such as humans. In both cases, data is most likely not discrete but continuous. This means that we have to estimate entropies, mutual information or conditional mutual information from continuous data without knowing the underlying probability distribution $\mu(x)$ (see Sect. 2.1). Approaches in this field can be divided into two major categories, *parametric* and *non-parametric* estimations. The first, *parametric* does require some a priori knowledge about the data. One example is the estimation of a probability mass functions by parametrised kernels, which are then used to calculate e.g. the entropy [13, 14].

Another commonly used technique, which is also used in the applications chapter (see Chap. 5) is the discretisation of the data into bins (see Sect. 2.1). This is illustrated with the example of estimating the mutual information $I(X; Y)$ (the example is taken from [3]). In this example, we have N bivariate measurements $z_i = (x_i, y_i)$, which are assumed to be iid (independent identically distributed) of a random variable $Z = (X, Y)$ and joint distribution $\mu(x, y)$. The marginal distributions of X and Y are $\mu(x) = \int dy \mu(x, y)$ and $\mu(y) = \int dx \mu(x, y)$. Mutual Information is then defined as

$$I(X; Y) = \int \int dxdy \mu(x, y) \ln \frac{\mu(x, y)}{\mu(x)\mu(y)}. \tag{2.103}$$

In the most common approach, known as the frequency based method or empirical method (see Sect. 2.1, the domains of the variables X and Y are partitioned into bins of finite size. Let $n_x(i)$ and $n_y(i)$ be the number of data points that fall into bin i-th and j-th bin of X and Y, and $n(i, j)$ is their intersection, then the distributions are approximated in the following way:

$$p_x(i) = \frac{n_x(i)}{N} \tag{2.104}$$

$$p_y(j) = \frac{n_y(j)}{N} \tag{2.105}$$

$$p(i, j) = \frac{n(i, j)}{N}. \tag{2.106}$$

Mutual information is then approximated using the definition given above (see Definition 2.6) in the following way:

$$I(X; Y) \approx I_{\text{binned}}(X; Y) = \sum_{i,j} p(i, j) \ln \frac{p(i, j)}{p_x(i)p_y(j)}. \tag{2.107}$$

This is an example for a parametrised approximation because the binning (number of bins, bin sizes and distribution of bins over the state space) are usually predetermined by the experimenter. The binning is chosen such that the underlying probability distribution is captured by the discretisation. Finding the best binning is not trivial.

We are interested in approximation methods that do not require a priori knowledge to find e.g. the optimal binning strategy. This is why we focus on non-parametric methods in this section. The reason is that we would like to use the methods on real data acquired e.g. from human motion. In such scenarios, the binning method discussed above suffers from the curse of dimensionality, i.e., the number of required calculations grows non-linearly with the number of random variables and bins, which makes them impracticable for high-dimensional systems. Hence, this section presents current developments in the field of entropy estimation on continuous spate spaces.

This section describes the estimation of entropies based on the k-nearest neighbour method as it was introduced by Kozachenko and Leonenko [15], but based on the more recent work by Lombardi and Pant [16] and Kraskov et al. [3], which both describe the original work by Kozachenko and Leonenko ([15] that is only available in Russian) in the same way.

The question is how to estimate the entropy of a (multivariate) random variable X in continuous space with a finite number of samples? The main idea is to estimate the probability mass around each sample by a local Gaussian approximation (see Sect. 2.1). The local approximation is based on the distance of the k-nearest neighbour.

Let $X \in \mathbb{R}^d$ be the random variable for which the entropy should be estimated and let the probability density over the state space be denoted by $\mu(x)$. The entropy of X is then given by

$$H(X) = - \int_{\mathbf{x}} dx \mu(x) \ln \mu(x), \tag{2.108}$$

where $\mathbf{x}$ is the support of $\mu(x)$.

Let x_i, $i = 1, 2, \ldots, N$ be our finite set of samples. A Monte-Carlo estimate of the entropy [17] is then given by

$$\hat{H}(X) = -\frac{1}{N} \sum_{i=1}^{N} \ln \mu(x). \tag{2.109}$$

Since $\mu(x)$ is unknown, we have to find an estimate $\hat{p}(x)$ to replace $\mu(x)$ in Eq. (2.109). The key idea that is used here (and in the two following sections) is to estimate $\hat{p}(x)$ through the k-nearest neighbour of x_i. The reasoning behind this form of estimation can be explained in the following way (Fig. 2.3).

Assume that there is an area in the state space which is very sparse, i.e., the distance between samples is considerably large. This is most likely an area of the state space in which samples are distributed very sparsely. This corresponds to a region with

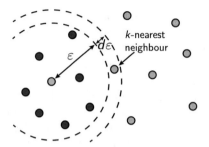

Fig. 2.3 Illustration of the ε-ball and the k-nearest neighbour. Redrawn from [16]. This image shows several data points $x_j \in \mathbb{R}^2$. The green data point is the data point x_i for which the $\hat{p}(x_i)$ is estimated by its ε-ball (depicted by the dashed lines). Exactly one data point is within the distance $[\varepsilon, \varepsilon + d\varepsilon]$ (light blue), $k-1$ data points are closer than the k-nearest neighbour (dark blue points), and the remaining data points (orange) are farther away

a low overall probability density, because a high probability density would lead to many samples in the corresponding region. This means that a large distance to the k-nearest neighbour relates to a low probability density and a small distance to the k-nearest neighbour related to a high probability density.

Hence, we consider the probability density $p_k(\varepsilon)$ of ε, where ε is the distance from x_i to its k-nearest neighbour. The probability $p_k(\varepsilon)d\varepsilon$ is the probability that exactly one point is in $[\varepsilon, \varepsilon + d\varepsilon]$, exactly $k-1$ points are closer than the k-nearest neighbour, and that the remaining points a farther than the k-nearest neighbour (see Fig. 2.3), i.e.,

$$p_k(\varepsilon)d\varepsilon = \binom{N-1}{1}\frac{dp_i(\varepsilon)}{d\varepsilon}d\varepsilon\binom{N-2}{k-1}(p_i(\varepsilon))^{k-1}(1-p_i(\varepsilon))^{N-k-1}, \quad (2.110)$$

where $p_i(\varepsilon)$ is the probability mass of an ε-ball centred at the sample point x_i [16]. The probability density $p_i(\varepsilon)$ is given by

$$p_i(\varepsilon) = \int_{\mathcal{B}(\varepsilon,x_i)} p(x)dx, \quad (2.111)$$

where $\mathcal{B}(\varepsilon, x_i)$ is the region inside the ε-ball, i.e., for which $\|x - x_i\| < \varepsilon$. The expectation of $\ln p_i(\varepsilon)$ can be calculated from the two previous equations:

$$\mathbb{E}(\ln p_i) = \int_0^\infty p_k(\varepsilon)d\varepsilon \ln p_i(\varepsilon) \quad (2.112)$$

$$= \psi(k) - \psi(N), \quad (2.113)$$

where $\psi(x)$ is the digamma function $\psi(x) = \Gamma(x)^{-1}d\Gamma(x)/dx$, which satisfies the recursion $\psi(x+1) = \psi(x) + 1$ and $\psi(1) = C$, and $C = 0.5772156\ldots$ is the

Euler-Mascheroni constant [3]. Assuming that $\mu(x_i)$ is constant over the entire ε-ball, this results in

$$p_i(\varepsilon) \approx c_d \varepsilon^d \mu(x_i), \tag{2.114}$$

where d is the dimension of x and c_d is the volume of the d-dimensional ε-ball. Note, that this is similar to the discrete binning estimation and different to approximating the probabilities locally by Gaussian functions (see Sect. 2.1). Two examples for Eq. (2.114) are the maximum norm (L_∞) $c_d = 1$ and for the Euclidean norm (L_2) $c_d = \pi^{d/2}/\Gamma(1 + d/2)$, where Γ is the Gamma function.

Taking the logarithm on both sides of Eq. (2.114) and using Eq. (2.113) in Eq. (2.109) results in:

$$\hat{H}(X) = -\frac{1}{N} \sum_{i=1}^{N} \ln \mu(x) \tag{2.115}$$

$$\approx -\frac{1}{N} \sum_{i=1}^{N} (\ln p_i(\varepsilon) - \ln c_d - \ln \varepsilon^d) \tag{2.116}$$

$$\approx -\frac{1}{N} \sum_{i=1}^{N} (\ln p_i(\varepsilon) - \ln c_d - \ln \varepsilon^d) \tag{2.117}$$

$$\approx -\mathbb{E}(\ln p_i(\varepsilon)) + \frac{1}{N} \sum_{i=1}^{N} \ln c_d + \frac{1}{N} \sum_{i=1}^{N} \ln \varepsilon^d \tag{2.118}$$

$$\approx \psi(N) - \psi(k) + \ln c_d + \frac{d}{N} \sum_{i=1}^{N} \ln \varepsilon_i, \tag{2.119}$$

where ε_i is the distance of the i-th sample to its k-nearest neighbours.

The assumption of uniformity in Eq. (2.114) is dropped in [16] and error estimations are given for the original and the improved entropy estimator. To summarise, the error depends on the error of estimating the underlying distribution by an ε-hypercube and it increases with the dimension of the data. Other variations are also discussed in e.g. [18–21].

2.8.1 Mutual Information Estimation on Continuous Data

The method for estimating mutual information on continuous state spaces follows the concept presented in the previous section on estimating entropy. The k-nearest neighbours are used to estimate the probability mass around data points. Kraskov et al. [3] present two methods to estimate mutual information which are based on the

work by [15] (see the previous section). The first method uses the following equality
(see Eq. 2.68):

$$I(X;Y) = H(X) + H(Y) - H(X,Y). \tag{2.120}$$

The two entropies $H(X)$ and $H(Y)$ can be calculated with Eq. (2.119). To estimate
the joint entropy $H(X,Y)$, we replace d by $d_Z = d_x + d_y$, c_d by $c_x + c_y$, and x by
$z = (x, y)$ in Eq. (2.114), which results in

$$\hat{H}(X,Y) = \psi(N) - \psi(k) + \ln(c_{d_x} + c_{d_y}) + \frac{d_x + d_y}{N} \sum_{i=1}^{N} \ln \varepsilon_i. \tag{2.121}$$

Mutual information is now estimated in the following way:

$$\hat{I}(X;Y) = \hat{H}(X) + \hat{H}(Y) - \hat{H}(X,Y) \tag{2.122}$$

which leads to

$$\hat{I}(X;Y) = \psi(N) - \psi(k) - \frac{1}{N} \sum_{i=1}^{N} \mathbb{E}[\psi(n_x(i) + 1) + \psi(n_y(i) - 1)], \tag{2.123}$$

$n_x(i)$ is the number of points closer to x_i than ε, and $n_y(i)$ is the number of points
closer to y_i than ε. We will use the following notation

$$\hat{I}(X;Y) = \psi(N) - \psi(k) - \langle \psi(n_x + 1) + \psi(n_y - 1) \rangle, \tag{2.124}$$

where

$$\langle \cdots \rangle = \frac{1}{N} \sum_{i=1}^{N} \mathbb{E}[\cdots (i)]. \tag{2.125}$$

It must be noted here, that the individual errors of the three entropy estimations
in Eq. (2.122) will generally not cancel each other out but rather add up. Furthermore, the Kozachenko-Leonenko estimator for entropy is only defined for marginal
distributions, and hence, is not correctly used for the estimation of $H(X,Y)$ by only
using one ε for both dimensions. This is why Kraskov et al. [3] proposed a second
method, which is introduced next.

The second method replaces $p_k(\varepsilon)$ by a two-dimensional density. An estimator for
more than two dimensions, i.e., $\hat{I}(X_1; X_2; \ldots; X_n)$ is discussed in [3] and the resulting
equation will be presented at the end of this section. For the second method, we now
have

$$p_k(\varepsilon_x, \varepsilon_y) = p_k^{(b)}(\varepsilon_x, \varepsilon_y) + p_k^{(c)}(\varepsilon_x, \varepsilon_y). \tag{2.126}$$

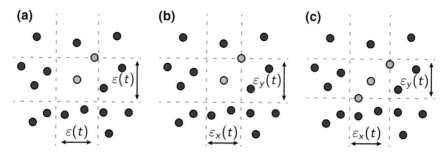

Fig. 2.4 *Visualisation of different types of ε-regions. Plot **a** shows $\varepsilon(t) = \max\{\varepsilon_x(t), \varepsilon_y(t)\}$, which lead to the first estimator for mutual information by Kraskov et al. [3]. Plots **b** and **c** show the two cases which are distinguished for the second estimator by Kraskov et al. [3]. The left-hand side shows case (**b**), in which one data point determines $\varepsilon_x(t)$ and $\varepsilon_y(t)$. The right-hand side shows case (**c**) in which $\varepsilon_x(t)$ and $\varepsilon_y(t)$ are determined by two different points. Images are redrawn from [3]*

The superscripts (*b*) and (*c*) refer to two different cases. The superscript (*b*) refers to the case in which a single point determines ε_x and ε_y while the superscript (*c*) refers to the case in which ε_x and ε_y are determined by two different points (see Fig. 2.4). The equations for $p_k^{(b)}(\varepsilon_x, \varepsilon_y)$ and $p_k^{(c)}(\varepsilon_x, \varepsilon_y)$ are given by:

$$p_k^{(b)}(\varepsilon_x, \varepsilon_y) = \binom{N-1}{k} \frac{d^2[q_i^k]}{d\varepsilon_x d\varepsilon_y}(1 - p_i)^{N-k-1} \tag{2.127}$$

$$p_k^{(c)}(\varepsilon_x, \varepsilon_y) = (k-1)\binom{N-1}{k} \frac{d^2[q_i^k]}{d\varepsilon_x d\varepsilon_y}(1 - p_i)^{N-k-1}, \tag{2.128}$$

where $q_i = q_i(\varepsilon_x, \varepsilon_y)$ is the mass of the rectangle of size $\varepsilon_x \times \varepsilon_y$ centred at (x_i, y_i), and p_i is defined as before as the mass of the square with size $\varepsilon = \max\{\varepsilon_x, \varepsilon_y\}$. The expected value of $\ln q_i$ is

$$\mathbb{E}(\ln q_i) = \int\int_0^\infty d\varepsilon_x d\varepsilon_y p_k(\varepsilon_x, \varepsilon_y) \ln q_i(\varepsilon_x, \varepsilon_y) \tag{2.129}$$

$$= \psi(k) - \frac{1}{k} - \psi(N). \tag{2.130}$$

This leads to the second estimator for bivariate mutual information

$$\hat{I}(X; Y) = \psi(k) - \frac{1}{k} - \langle \psi(n_x) + \psi(n_y)\rangle + \psi(N), \tag{2.131}$$

where $\langle \cdots \rangle$, n_x, and n_y are defined as before (see above). Without further explanation, the estimator for multivariate mutual information is given by

$$\hat{I}(X_1; X_2; \ldots; X_n) = \psi(k) - \frac{n-1}{k} + (n-1)\psi(N) - \left\langle \sum_{i=1}^{n} \psi(n_{x_i}) \right\rangle. \quad (2.132)$$

For details, the reader is referred to [3].

To conclude this section, the method of estimating mutual information on continuous state spaces [3] that is used in this book is based on the method of estimating entropies proposed by [15]. The probability mass function is estimated by counting the number of points within an ε-region of a ball, i.e., by looking at the distance of the k-nearest neighbour. Special attention must be given in case of the mutual information on how the ε-hypercube is generated, which lead to the second estimator.

2.8.2 Conditional Mutual Information Estimation on Continuous Data

The estimator for conditional mutual information on continuous state spaces by Frenzel and Pompe [4] is based on the estimator of differential entropy by Kozachenko and Leonenko [15] that was discussed above (see Sect. 2.119).

Conditional mutual information can be translated into sums of entropies (see Definition 2.7). Applying the chain rule for entropies (see Eq. 2.24), we get

$$I(X; Y|Z) = H(X|Z) - H(X|Y, Z) \quad (2.133)$$
$$= H(X, Z) + H(Y, Z) - H(X, Y, Z) - H(X). \quad (2.134)$$

The estimator for the conditional mutual information by Frenzel and Pompe is then given by

$$\hat{I}(X; Y|Z) = \hat{H}(X, Z) + \hat{H}(Y, Z) - \hat{H}(X, Y, Z) - \hat{H}(X). \quad (2.135)$$

Hence, this estimator for conditional mutual information builds upon the previously introduced estimator for entropy (see Eq. (2.109)) and the generalisation introduced by Kraskov et al. [3] for their first estimator of mutual information (see Eq. (2.121)). The same methodology that was used by Kraskov et al. is used by Frenzel and Pompe to generalise the entropy estimation to three random variables. The authors use the maximum norm, i.e., $\| \cdot \| = \max\{\| \cdot \|_x, \| \cdot \|_y, \| \cdot \|_z\}$ (analogous for one and two random variables). This means that this estimator can suffer from the difficulties that were discussed above as the motivation to derive the second estimator for mutual information by Kraskov et al.

Let n_{z_i}, n_{xz_i}, and n_{yz_i} be defined as the number of points that fall in the ε-balls which are centred around z_i, (x_i, z_i), and (y_i, z_i). Then, the estimator for conditional mutual information on continuous state spaces is given by:

$$\hat{I}(X; Y|Z) = \langle \psi(n_{xz_i}) + \psi(n_{yz_i}) + \psi(n_{z_i}) \rangle - \psi(k-1), \quad (2.136)$$

where $\psi(x)$ and $\langle \cdot \rangle$ are defined as above. Equation (2.136) reduces to Eq. (2.122) if Z is disregarded, i.e., $Z = \emptyset$ and $I(X; Z) = I(X; Y|\emptyset)$.

2.9 Conclusion

The quantifications that are presented in the following chapters rely on information theory. Therefore, this chapter presented the information-theoretic foundation for the main results of this book. Most of this chapter and all of the presented quantifications are given for discrete state spaces. The reason is that this allows us to focus on the development of the theory. The last section of this chapter presented estimators for entropy, mutual information, and conditional mutual information that operate on continuous data. These are used in two instances in this book. First, in the applications chapter (see Chap. 5) they are applied to data from hopping models to compare the results obtained from the two different estimation methods (discrete vs. continuous) that were discussed in this chapter. We show that the results are qualitatively equivalent. This allows the conclusion that several of the quantifications presented in this book can be applied to real-world data, e.g. motion capturing data from human motion.

References

1. Shannon CE (1948) A mathematical theory of communication. Bell Syst Techn J 27:379–423
2. Cover TM, Thomas JA (2006) Elements of information theory, vol 2. Wiley, Hoboken, New Jersey, USA
3. Kraskov A, Stögbauer H, Grassberger P (2004) Estimating mutual information. Phys Rev E 69:066138
4. Frenzel S, Pompe B (2007) Partial mutual information for coupling analysis of multivariate time series. Phys Rev Lett 99:204101
5. Blitzstein JK, Hwang J (2014) Introduction to probability. CRC Press
6. Klenke A (2014) Probability theory. Springer
7. (2018) Law of large numbers. Encyclopedia of Mathematics. http://www.encyclopediaofmath.org/index.php?title=Law_of_large_numbers&oldid=26552
8. Lindsay BG (1995) mixture models: theory, geometry and applications. NSF-CBMS Conference series in Probability and Statistics, Penn. State University
9. Marin JM, Mengersen KL, Robert CP (2005) Handbook of statistics: volume 25, Elsevier, chap Bayesian modelling and inference on mixtures of distributions
10. Pearl J (2000) Causality: models, reasoning and inference. Cambridge University Press
11. Ay N, Polani D (2008) Information flows in causal networks. Adv Complex Syst 11(1):17–41
12. Ay N, Zahedi K (2014) On the causal structure of the sensorimotor loop. In: Prokopenko M (ed) Guided self-organization: inception, emergence, complexity and computation, vol 9. Springer, pp 261–294
13. Moon YI, Rajagopalan B, Lall U (1995) Estimation of mutual information using kernel density estimators. Phys Rev E 52:2318–2321
14. Steuer R, Kurths J, Daub CO, Weise J, Selbig J (2002) The mutual information: Detecting and evaluating dependencies between variables. Bioinformatics 18(2):S231–S240
15. Kozachenko LF, Leonenko NN (1987) Sample estimate of the entropy of a random vector. Probl Inf Transm 23(1):95–101
16. Lombardi D, Pant S (2016) Nonparametric k-nearest-neighbor entropy estimator. Phys Rev E 93(1):013310
17. Ahmad I, Lin PE (1976) A nonparametric estimation of the entropy for absolutely continuous distributions (corresp.). IEEE Trans Inform Theory 22(3):372–375

18. Tsybakov AB, van der Meulen EC (1996) Root-n consistent estimators of entropy for densities with unbounded support. Scand J Stat 23(1):75–83
19. Singh H, Misra N, Hnizdo V, Fedorowicz A, Demchuk E (2003) Nearest neighbor estimates of entropy. Am J Math Manag Sci 23(3–4):301–321
20. Gray AG, Moore AW (2003) Nonparametric density estimation: toward computational tractability. In: SDM
21. Mnatsakanov RM, Misra N, Li S, Harner EJ (2008) K n-nearest neighbor estimators of entropy. Math Methods Stat 17(3):261–277

Chapter 3
A Theory of Morphological Intelligence

Intelligence is making a difficult problem easy

David Krakauer

Pfeifer and Iida [1] and Paul [2] state that "*One problem with the concept of morphological computation is that while intuitively plausible, it has defied serious quantification efforts.*" Paul [2] adds "*We would like to be able to ask: How much computation is actually being done?*" In the context of this work, we would rephrase this question in the following way. Given a time series of a recorded behaviour (e.g. 3D motion capturing data), we would like to post hoc determine how much of the behaviour was controlled by the brain and how much of it resulted from the exploitation of the body-environment dynamics? This is the question of quantifying morphological intelligence from observational data.

Hence, this book presents a quantitative approach to understanding and modelling morphological intelligence.

This chapter is organised in the following way. It will start with an overview of related work. It must be noted here, that the related work was formalised with respect to morphological computation, which is understood as a subset of morphological intelligence in the context of this book. The two main approaches to formalising morphological computation can be categorised into a dynamical systems and an information-theoretic approach. Both approaches are discussed in more detail. The discussion of related work is then followed by the presentation of four concepts to quantify different aspects of morphological intelligence. The fifth concept is a proposal to quantify morphological intelligence as it was introduced in the previous chapter (see Definition 1.1).

© Springer Nature Switzerland AG 2019
K. Ghazi-Zahedi, *Morphological Intelligence*,
https://doi.org/10.1007/978-3-030-20621-5_3

3.1 Related Work on Formalising Morphological Computation

There are basically two different streams of formalising morphological intelligence or aspects of it, which can be divided into a dynamical systems and an information-theoretic approach. The two approaches do not stand in opposition, but should rather be seen as complementary [3]. The first approach [4, 5] models morphological computation in the context of reservoir computing [6, 7]. This means that the body is understood as a physical reservoir computer and the controller or brain harnesses the body dynamics. Two examples are the spine-driven robot (see Sect. 1.1.2 and [8]), which uses the spine dynamics as part of its controller and the artificial octopus arm that uses its body dynamics for computation (see Sect. 1.1.1 and [9]). Within this first approach, there are also several works which discuss the importance of tight body-brain-environment coupling, of which the following are just a few examples [10–16]. The dynamical systems approach will be discussed in more detail in Sect. 3.1.1 below. Although very intuitive and compelling, this approach does not allow to quantify how the body reduces the computational burden for the brain. As stated above, this book follows a quantitative approach to understanding morphological intelligence, which is the motivation for the second, information-theoretic approach [17]. The guiding idea is to model the sensorimotor loop as a causal graph (see Sect. 3.2 [18]) and, based on that, ask how much internal processes contributed to an observed behaviour as opposed to external processes (body-environment interactions). Information-theoretic measures have been successfully applied to quantify morphological computation in soft robotics (see Sect. 5.1 [19]) and muscle models (see Sect. 5.2 and [20]).

Next, we will first present a dynamical systems perspective on morphological computation and morphological control, before related work in the context of information theory is presented.

3.1.1 Dynamical Systems Approach to Formalising Morphological Intelligence

"*A dynamical system is a system that evolves in time through the iterated application of an underlying dynamical rule*" [21]. Similar definitions can be found throughout literature (e.g. [22–24]), yet, the remainder of this section will use the definitions as they are presented by Jost [21].

The first required definition describes the underlying dynamical rule that was mentioned in the quote above.

Definition 3.1 (*Flow*) A flow (semiflow) is a family

$$F_t : X \to X \tag{3.1}$$

of maps of a set X (state or phase space) into itself, for $t \in \mathbb{R}$ ($t \geq 0$), satisfying

1. $F_0 = Id$
2. $F_{t+s} = F_t \circ F_s$ for all $t, s \in \mathbb{R}$ ($t, s \geq 0$) ((semi) group property).

The parameter t is referred to as time. In the context of dynamical systems (and the related work that will follow below), we are interested in the evolution of the system over time, given an initial condition x_0:

$$x(t) := F_t x_0. \tag{3.2}$$

In the following, we will define the terms *orbit, transient, attractor,* and *basin of attraction* because they are necessary to understand the dynamical systems approach to formalising morphological computation.

The *orbit* of a dynamical system is the collection of points that describes the evolution of the system over time, i.e., $\{x(t) : t \in \mathbb{R}(t \geq 0)\}$ is the orbit of x_0.

The initial condition x_0 is called a *fix-point* or *stationary point*, if

$$x(t) = x_0 \text{ for all } t. \tag{3.3}$$

The general case of a fix-point is a periodic orbit, i.e.,

$$x(t + \omega) = x(t) \text{ for some } \omega \geq 0 \text{ and all } t \tag{3.4}$$

In the discrete case, i.e., in which the time t is discrete, denoted by $n \in \mathbb{N}$, we consider iterated maps, which are given by:

$$F : X \rightarrow X \tag{3.5}$$
$$x_{n+1} = F(x_n) \text{ for } n \in \mathbb{N}. \tag{3.6}$$

As stated by Jost [21] and Strogatz [22], one of the most fundamental concepts of dynamical systems (*attractor*) does not have a universally accepted definition. We present the definition by [21] below. Stated colloquially, an attractor is a periodic orbit with initial conditions that are not included in the periodic orbit, i.e., which converge towards itself.

Definition 3.2 (*Attractor*) A compact set A in a topological space X is called an attractor for the continuous map $F : X \rightarrow X$ if there exists a neighbourhood U of A with

$$F(U) \subset U \tag{3.7}$$

and

$$A = \bigcap_{n \in \mathbb{N}} F^n(U). \tag{3.8}$$

Fig. 3.1 *Visualisation of transients, attractors, and basin of attraction.* This image is redrawn from [4]. The plot on the left-hand side shows a phase space with two attractors (limit cycles) and their corresponding basins of attraction. The plot in the centre shows that the attractor on the lower half has changed its shape (compared with the plot on the left-hand side). The plot on the right-hand side shows that a new attractor has emerged (compared to the plot shown in the left-hand side). The plots are discussed in context of body-environment interactions in the text below

Attractors can also be defined similarly for time-continuous systems.

The final definition is *basin of attraction*, which is the set of initial conditions that end up in a particular attractor.

Definition 3.3 (*Basin of Attraction*) The largest open set U that satisfy the conditions of Definition 3.2 is called the basin of attraction of A.

To summarise the previously given definitions and concepts in colloquial terms: Given an initial condition x_0, the flow (continuous time) or iterated maps (discrete time) describe the evolution of the system over time. The set of states that originate in x_0 are called the orbit of x_0. A state x that is its own orbit is called a fix-point, and a set of points which are their own orbit is a periodic orbit. A fix-pint or periodic orbit are called an attractor, if they are surrounded by initial points which converge towards them. The set of all points which converge towards an attractor is the basin of attraction.

Orbit, attractor, and basin of attraction are visualised in Fig. 3.1. Figure 3.1A shows two attractors (orange and blue ellipsoids), several orbits (lines with arrowheads) and two basins of attraction, separated by the red line.

It must be noted here that this is a very reduced presentation of dynamical systems that omits many important concepts, such as stability, instability, chaos, etc. We chose to present only the most fundamental concepts because they are sufficient to understand the remainder of this section. For more detail, the reader is referred to [21–23].

We can now describe Puppy's interaction with the environment ([25–27], and Sect. 1.1.1) in the context of dynamical systems.

Puppy is an under-actuated four-legged walking machine with springs in the second joint of each leg. It is driven by a simple open-loop controller which drives the motors with oscillatory signals. The overall behaviour pattern of Puppy, i.e., the frequency of the hopping movements, results from an interaction of the open-loop controller, the leg-morphology (e.g. springs and feet material), and the environment (e.g. ground texture).

As discussed by Pfeifer and Bongard [10], the behaviour of the system (Puppy) is best described by an attractor landscape. The attractor landscape is formed by the interaction of the three components, which means that the attractor landscape, i.e., the attractors and their basins of attraction depend on controller, morphology, and environment. As an example, if Puppy is set onto the ground with a particular texture, e.g. a carpet, it will start in a different basin of attraction compared to e.g. concrete. As a result, the hopping movement will differ. If we now change the material of the feet or the spring constants, this can be understood as reshaping the attractor landscape. As a result Puppy's behaviour might converge to a different attractor when placed on the same ground. This is shown schematically in Fig. 3.1b. It is possible that instead of just reshaping the landscape, entirely new attractors with a new basin of attractions occur as a result of changes to the systems (see Fig. 3.1c).

Given this example, we can now focus on the approach that formalises morphological computation in the context of dynamical systems. For this purpose, it is important to restate the definition of morphological computation that is used in the dynamical systems approach. Füchslin et al. [5] write

> In 2007, a workshop at the first International Conference on Morphological Computing in Venice, Italy, led by Norman Packard, informally defined morphological computing as "any process that (a) serves for a computational purpose, (b) has clearly assignable input and output states and (c) is programmable, where 'programmable' is understood in the broad sense that a programmer can vary the behaviour of the system by varying a set of parameters."

To summarise this quote, morphological computation requires a purpose, must have clear input and output states and must be programmable in some sense. As we will discuss below, point (c) refers to reshaping of the attractor landscape in the way it was discussed for Puppy above.

Based on this definition, the authors [5] formalise this notion of morphological computation in the following way. A *programmable dynamical system* is defined as the triple $\Pi = (S, M, (f_j)_{j \in J})$, where S are the states, $(M, <)$ is a partially ordered monoid, and $f : S \times M \to S$ is a transition function. This system is called a programmable dynamical system because for each $j \in J$, the triple (S, M, f_j) is a dynamical system and $j \in J$ is the *program*. A programmable dynamical system will be related to the physical dynamics of a morphology (see Fig. 3.1), and later, to physical reservoir computing. But first, functions are needed that translate inputs to states of the dynamical and states of the dynamical systems to outputs. Füchslin et al. [5] call these functions *computational interpretations* and define them as a map from natural numbers to initial states and as a map from terminal states (attractors) to natural numbers, denoted by $C_\Gamma : \mathbb{N} \to S$ and $D_\Gamma : \Theta(\Gamma) \to \mathbb{N}$. For more details, the reader is referred to [5]. Morphological computation can now be described by the commutative diagram shown in Fig. 3.2 (which is very similar to the commutative diagram found in [28], see Fig. 1.6).

The diagram that is displayed in Fig. 3.2 should be understood in the following way. The function $\psi : \mathbb{N} \to \mathbb{N}$ (abstract calculation, e.g. $2 + 2 = 4$) is computed

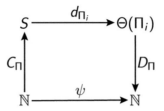

Fig. 3.2 *Commuting diagram of morphological computation.* This diagram is redrawn from [5]. It shows how morphological computation is understood in the context of dynamical systems. In their example, a natural number is encoded into an initial condition S of the programmable dynamical system using the encoding function C_Π. The physical dynamics of the system, denoted by d_{Π_i}, transforms the initial state S into a terminal state $\Theta(\Pi_i)$, which is decoded back into a natural number with the decoding function D_Π. The system is said to be computing if this diagram applies. Not that for morphological computation d_{Π_i} are dynamical processes that happen in the body, i.e., are purely physical processes. For detail, please see the text below

by a programmable dynamical system by first encoding the natural number into an initial state. The encoding function is denoted by $C_\Gamma : \mathbb{N} \to S$. In the plots shown in Fig. 3.1, this could refer to initialising the system in the orange basin of attraction. The dynamical system then converges to its terminal state $\Theta(\Gamma)$, denoted by d_{Π_i}. This refers to the fundamental physical processes that were described in the theory of computation by Horsman et al. [28]. The terminal state (attractor) is decoded back into a natural number by D_Π. This naturally maps to the concept of reservoir computing [6, 7, 29].

In general, reservoir computing refers to recurrent neural networks, whose dynamics are driven by an input layer that connects input neurons to all or a subset of the reservoir nodes. A linear output layer transforms the reservoir's state to output values. Hence, the input layer refers to the map C_Γ and the output layer analogously to the map D_Γ. The only difference is that in reservoir computing, the reservoir does usually not converge before the output values are presented. To conclude, the theory of physical computation by Horsman et al. [28] is translated to dynamical systems to model morphological computation in systems such as Puppy, the spine-driven robot and the octopus arm that were discussed earlier.

Rückert and Neumann [30] connect morphological computation to the theory of probabilistic optimal control. They investigate how much computation "*can be absorbed by the morphology of a robot*", by changing the morphology of a simplified simulated arm and computing the optimal control strategy for it. In this context, optimality refers to a reduction of complexity while ensuring the same or better performance. Hence, the complexities of optimal controllers are compared to estimate how much the morphology reduced the computational demand. This is already very close to our definition of morphological intelligence (see Definition 1.1).

Corucci et al. [31] use a similar approach. They compare the entropy of controllers for various morphologies to estimate how much the morphology contributed to the behaviour. As in the previous examples, morphological intelligence is measured indirectly by comparison of control complexities.

Cheap control [10, 32] and morphological intelligence are, from our perspective, heavily intertwined. Morphological intelligence (see Definition 1.1) is the reduction of computational cost as a result of the exploitation of the body and its interaction with the environment. Hence, a system that shows morphological intelligence is necessarily also cheaply designed as described by [10]. On the other side, in order to have a cheap controller, the system must exploit the embodiment, i.e., have morphological intelligence. Hence, although most of the examples given in this section focus on control aspects, they are still related in the sense that they quantify control effort, and therefore also indirectly quantify morphological intelligence. The difference to the work presented in this book is that we focus on directly measuring the contribution of the morphology to cognition and intelligence.

Polani [33] minimizes the controller complexity in a reinforcement learning setting. In particular, the authors ask the question, what is the minimally required information for the policy or brain of an embodied system, such that it can still fulfil a given task. This information is referred to as the *relevant information*. The principle is formulated in the following equation:

$$\min_{p(a|s)} \left(I(S; A) - \beta \mathbb{E}\{Q(s, a)\} \right), \tag{3.9}$$

where $I(S; A)$ is the complexity of the policy and $\mathbb{E}\{Q(s, a)\}$ is the expected reward in a reinforcement learning setting. Hence, in order to minimise Eq. (3.9), the expected reward must be maximised and the complexity of the controller must be minimised. Minimising the complexity of the controller, as argued before, implies morphological intelligence.

The final work discussed in this section is the work by Haeufle et al. [34], who quantify the control effort in muscles and motor models. A simplified system, which is equipped with a single linear actuator (muscle or motor), has to jump up and down with constant hopping height. The control effort of several different muscles and motor actuators were evaluated (a subset is also used in the applications section in this work, see Sect. 5.2) which means that the entropy of the sensor states is calculated, i.e., $H(S)$, where S is the discretised sensor state. The authors then ask the question of how far the number of bins in the discretisation can be reduced without losing the system's hopping behaviour. This is related to asking what is the minimal amount of information that the system has to acquire through its sensors to maintain its behaviour, which is again equivalent (in an information-theoretic setting) to asking what is the minimal complexity of the required controller.

This concludes the section on related work from two different approaches, namely dynamical systems and information theory. The next section will present the causal model of the sensorimotor loop, which is the foundation for most of the concepts described in this chapter.

3.2   Causal Model of the Sensorimotor Loop

The majority of the quantifications that are discussed in the following sections rely on a causal model of the sensorimotor loop, which is presented in this section. The model of the sensorimotor loop used in this section is closely related to the models discussed by [3, 35, 36]. A detailed discussion can also be found in [37].

In this work, we follow von Förster [38, 39] and define cognition as the very basic process that transforms sensory data into motor commands, using a form of internal (non-symbolic) representation. Hence, a cognitive system must at least have a sensory system (or just sensors for short) and a motor system (or actuators for short). In accordance with [40–42] and more recent work by [10, 32], it follows that a cognitive system is embedded in an environment and cannot be understood or modelled detached from it.

The remainder of this work is based on the assumption that there is a canonical way to separate a cognitive system into its components, namely, brain, sensors, actuators, and body. We further assume, that there is a clear system-environment separation. We are fully aware that the body-environment separation is a very difficult and yet unsolved question for biological systems (see e.g. [43], for a discussion). This holds even more for the distinction between body and brain. The human vision system is an example of the problem of brain-body distinction. One can make the case that the visual cortex is not the first part of the brain that processes visual stimuli, but that processing already occurs in the optic chiasma (where neural pathways from the left and right eye cross). One can also make the case that the retina itself is processing information (see Sect. 1.1.3). The retina contains approximately 130 million retinal receptors [44] but in comparison only approximately 1.2 million optic nerve fibres, which means that a lot of pre-processing must already happen in the retina. Early studies have also shown that the retina contains specialised cells for orientation, edge, and uniformity detection [45].

Yet, in order to derive a quantification for morphological intelligence, we have to make a distinction between body and brain, which also includes a distinction between body, brain, sensors, and actuators. The quantification given in the next sections, therefore, ask the following question: Given a particular modularisation of a cognitive system, how much of the observed behaviour resulted from external dynamics as opposed to internal dynamics. The following paragraphs will clarify, what we mean by external and internal, but first, we will describe the interaction of the components of a sensorimotor loop (see Fig. 3.3).

The controller or brain sends signals to the actuators. In the case of a biological system, these could e.g. be the action potentials that the nerves transport from the motor cortex to the muscles. The actuators influence the environment and can also be influenced by the environment (see Fig. 3.3). The first part is obvious, as the muscles in an arm and hand can be used to lift and reposition objects, but the objects also influence the state of the actuators, which is the arrow that connects the environment back to the actuators. In the example of the object that is lifted, the object's shape also limits the movement of the fingers. For this reason, we prefer the notion of the system's

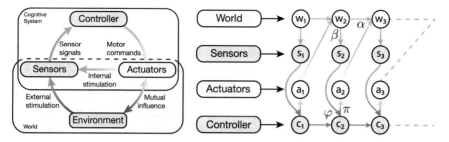

Fig. 3.3 *Schematics and causal graph of the sensorimotor loop.* The figure on the left-hand side shows the conceptual understanding of the sensorimotor loop. A cognitive system consists of a controller, a sensor and actuator system, and a body which is situated in an environment. The basic understanding is that the controller sends signals to the actuators which affect the environment. Information about the environment and internal states are sensed by the sensors, and the loop is closed when this information is passed to the controller. The figure on the right-hand side shows the representation of the sensorimotor loop as a causal graph. Here w_t represents the world state at time t. The world is everything that is physical, i.e. the environment and the morphology. The variables s_t and a_t are the signals provided by the sensors or passed to the actuators, respectively. They are not to be mistaken with the sensors and actuators, which are part of the morphology, and hence, part of the world

Umwelt [16, 46–48] over the term environment, because it captures the part of the system's environment that can be affected by the system, and which itself affects the system. The state of the actuators and the *Umwelt* are not directly accessible by the cognitive system, but the loop is closed as information about both, the *Umwelt* and the actuators are provided to the controller by the system's sensors. Although not shown in the schematics, this also includes other internal (proprioceptive) information, e.g. energy. Revisiting the human vision system as an example, we would state that the retina is part of the sensory system and that the action potentials that leave the retina are the sensor signals that are sent to the brain. This point will be revisited below when the causal diagram is introduced. This concludes the sensorimotor loop as it is commonly used in the embodied artificial intelligence community (see e.g. [3]).

We extend the sensorimotor loop by the notion of *world*. The *world* captures the system's morphology and the system's *Umwelt*. In more colloquial terms, the world is everything that an external observer can observer about the system and its *Umwelt*. This also means that the body is in some regard external to the system. This is analogous to the agent-environment distinction made in the context of reinforcement learning [49], where the environment is defined as everything that cannot be changed arbitrarily by the agent. Consequently, in our notion of the sensorimotor loop, we consider the brain, sensors and actuators, or more precisely, the brain state, the signals sent from the sensor to the brain and the signals sent from the brain to the actuators as *intrinsic* to the system. The world state, which contains the state of the system's body and *Umwelt* are considered *extrinsic* to the system.

The distinction between intrinsic and extrinsic is also captured in the representation of the sensorimotor loop as a causal or Bayesian graph (see Fig. 3.3, right-hand side). Random variables C, A, W, and S refer to the controller, actuator signals, world and sensor signals. The directed edges reflect causal dependencies between the random variables (see also [35, 47, 50]). Everything that is extrinsic is captured in the variable W, whereas S, C, and A are intrinsic to the system. The random variables S and A are not to be mistaken with the sensors and actuators. The variable S is the output of the sensors. In the example given above, these were described as the action potential sent from the retina to the brain. The robotic equivalent could the pixel matrix that is provided by a camera sensor. The random variable A is the input that the actuators take. In the example given above, this was described as the action potentials that the brain sends to the muscles. In an artificial system, this could be the numerical value that is given to a motor controller, which then converts it into currents to control a motor.

We can now develop a formal description of the sensorimotor loop. For simplicity, we only consider systems with discrete state space. For a more general discussion of maps, kernels, and probability distribution, the reader is referred to literature on probability theory (see e.g. [51]).

The set of all world states is denoted by $\mathcal{W}$. Denoting the set of sensor states by $\mathcal{S}$, we can consider the sensor to be an information transmission channel from $\mathcal{W}$ to $\mathcal{S}$ as it is defined within information theory [52]. Given a world state $w \in \mathcal{W}$, the response of the sensor can be characterized by a probability distribution $\beta(s|w)$ of possible sensor states $s \in \mathcal{S}$ as a result of w. For instance, if the sensor is noisy, then its response will not be uniquely determined. If the sensor is noiseless, that is, deterministic, then there will be only one sensor state as a possible response to the world state w. The response of the sensor given w can be described in terms of a Markov kernel

$$\beta: \mathcal{W} \longrightarrow \Delta_{\mathcal{S}}, \tag{3.10}$$

where $\Delta_{\mathcal{S}}$ denotes the set of probability distributions on the set $\mathcal{S}$ of sensor states. We call $\beta(s|w)$ the *sensor map*. Markov kernels are closely related to conditional probabilities, which would justify the notation $p(s|w)$ instead of $\beta(s|w)$. Following Pearl's concept of causal networks [53], the Markov kernels formalise the mechanisms of the sensorimotor loop. The difference between a conditional probabilities distributions and a mechanism is that a conditional probability distribution can be calculated or sampled between any two random variables while a mechanism refers to the causal dependence between the two variables in question. In all the models presented within this book, causal dependencies are visualised by arrows between the two corresponding variables and denoted by Greek letters. Hence, mechanisms refer to random variables with a causal dependence (and an arrow between them), while lower-case Latin letters refer to conditional probability distributions between two random variables without a direct causal link.

Once the mechanisms are defined, they generate distributions so that we can also compute the conditional distributions from them. For instance, one can compute the conditional distribution $p(s|w) = p(s,w)/p(w)$ and compare this with the mechanism $\beta(s|w)$. Clearly, they coincide whenever $p(w) > 0$, which we consider as a consistency property. However, there is an important difference, reflecting the fact that β is a mechanism: it is defined for *all* w. If the behaviour of an agent is restricted to only a few world states, then the conditional distribution $p(s|w)$ will be defined only for these world states.

It is now straightforward to describe a corresponding formalisation of the remaining components of the sensorimotor loop. The next step is the internal representation and memory of the agent, denoted by C in reference to the controller state. The corresponding mechanism φ is defined in the following way:

$$\varphi: \mathcal{S} \times \mathcal{C} \longrightarrow \Delta_{\mathcal{C}}. \tag{3.11}$$

It will be used in the following form $\varphi(c'|s, c)$ and called the *memory map* throughout this book. Systems that are equipped with a memory map are called *non-reactive systems*. Reactive systems do not have a memory map, and hence, no internal state C. This means, that a reactive policy π (described next) acts on the sensor information S instead of the internal state C (compare Fig. 3.3 with Fig. 3.4).

The agent can influence the world by means of actions. A reactive *policy* generates actions from the sensor state and is defined as

$$\pi: \mathcal{S} \longrightarrow \Delta_{\mathcal{A}}. \tag{3.12}$$

It is denoted by $\pi(a|s)$. A non-reactive policy generates actions from the internal state C. It is defined as

$$\pi: \mathcal{C} \longrightarrow \Delta_{\mathcal{A}} \tag{3.13}$$

and denoted by $\pi(a|c)$. We use the same symbol π in both cases as the type of policy (reactive vs. non-reactive) can be determined by the set of random variables it operates on. A different non-reactive policy, that also takes a desired next sensors state $\tilde{S}$ into account is discussed in the context of deliberate decision making [37].

Finally, we consider the change of the world state from w to w' in the context of an actuator state a as a channel, denoted by α, which assigns a distribution $\alpha(w'|w, a)$ to w, a. With the set $\Delta_{\mathcal{W}}$ of probability distributions on $\mathcal{W}$, we have

$$\alpha: \mathcal{W} \times \mathcal{A} \longrightarrow \Delta_{\mathcal{W}}. \tag{3.14}$$

We refer to $\alpha(w'|w, a)$ as *world dynamics kernel*. Similar causal structures, involving Markov kernels, have been considered in the general context of control and systems theory [54], as well as in the context of population dynamics [55].

To summarise the previous paragraphs, the level of abstraction that is used in the context of this work to describe the sensorimotor loop of an embodied agent is given by the following set of Markov kernels for a non-reactive system (see Fig. 3.3)

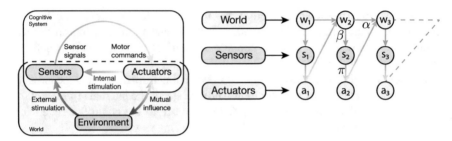

Fig. 3.4 *Causal diagram of the reactive system's sensorimotor loop.* In the context of this work, a reactive system is defined by direct coupling of the sensors and actuators. There is no form of memory or controller state present in the system

$$\beta: \mathcal{W} \longrightarrow \Delta_{\mathcal{S}} \qquad\qquad [\beta(s|w)] \qquad\qquad (3.15)$$

$$\varphi: \mathcal{C} \times \mathcal{S} \longrightarrow \Delta_{\mathcal{C}} \qquad\qquad [\varphi(c'|c, w)] \qquad\qquad (3.16)$$

$$\pi: \mathcal{C} \longrightarrow \Delta_{\mathcal{A}} \qquad\qquad [\pi(a|c)] \qquad\qquad (3.17)$$

$$\alpha: \mathcal{W} \times \mathcal{A} \longrightarrow \Delta_{\mathcal{W}} \qquad\qquad [\alpha(w'|w, a)] \qquad\qquad (3.18)$$

and the following set of Markov kernels for a reactive system (see Fig. 3.4)

$$\beta: \mathcal{W} \longrightarrow \Delta_{\mathcal{S}} \qquad\qquad [\beta(s|w)] \qquad\qquad (3.19)$$

$$\pi: \mathcal{S} \longrightarrow \Delta_{\mathcal{A}} \qquad\qquad [\pi(a|s)] \qquad\qquad (3.20)$$

$$\alpha: \mathcal{W} \times \mathcal{A} \longrightarrow \Delta_{\mathcal{W}} \qquad\qquad [\alpha(w'|w, a)]. \qquad\qquad (3.21)$$

In the remainder of this book, we will often abbreviate the random variables for better comprehension, as all measures consider random variables of consecutive time indices. Therefore, we use the following notation. Random variables without any time index refer to time index t and hyphened variables to time index $t + 1$. The two variables W, W' refer to W_t and W_{t+1}. In rare cases, we will also require time index $t - 1$, which is denoted by e.g. $W^{\ddagger}$ for W_{t-1}.

This concludes the discussion of the sensorimotor loop. The next section will present the first concept of quantifying morphological intelligence.

3.3   Concept One: Quantifying Morphological Intelligence Based on the Effect of the Action on the World

We will derive the first concept based on the causal diagram of the sensorimotor loop as it was discussed in the previous section (see Sect. 3.2). The starting point for our considerations is the world dynamics kernel $\alpha(w'|w, a)$, which captures the influence of the current action A and the current world state W on the next world state W'. Let us now assume that the current action A has an influence on the next

(a) Full Model: This causal diagram is the one-step version of the sensorimotor loop for reactive systems shown in Figure 3.4.

(b) Model in which the current action A has no influence on the next world state W'

Fig. 3.5 *Visualisation of MI$_A$.* This figure shows the two models of the sensorimotor loop that are compared in the measure

world state W', which means that the behaviour of our system is fully determined by the current world state W. This corresponds to the situation in which the system's behaviour is completely determined by the body dynamics, i.e., the interaction of the morphology and environment. One example for such a system, the Passive Dynamic Walker [56–58], was discussed in the first chapter of this book (see Chap. 1).

Formally, this means that the world dynamics kernel $\alpha(w'|w, a)$ reduces to $\hat{\alpha}(w'|w)$ (see Fig. 3.5b), which can be calculated in the following way

$$\hat{\alpha}(w'|w) = \sum_a \alpha(w'|w, a)p(a|w). \tag{3.22}$$

We will refer to $\hat{\alpha}(w'|w)$ as the assumption that the system's behaviour is independent of the system's action. Every difference from this assumption means that action A had an influence. In other words, $\alpha(w'|w, a)$ differs from our assumption $\hat{\alpha}(w'|w)$ if action A had an influence on the next world state W'. In this concept, this reduces the amount of morphological intelligence that was present in the observed behaviour.

The discrepancy of these two distributions can be measured with the Kullback-Leibler divergence $D_{\mathrm{KL}}(\alpha(w'|w, a)||p(w'|w))$, which leads to the following definition:

Definition 3.4 (MI$_A$) Let the random variables A, W, W' denote the current action, the current world state and the next world state of an embodied agent, as illustrated in the causal diagram shown in Fig. 3.5a. The effect of the current action on the behaviour is then defined in the following way:

$$\mathrm{MI}_A = D(p(w'|w, a)||p(w'|w)) \tag{3.23}$$

MI$_A$ is available as part of *gomi* (see Sects. A.2.2 and A.3.2).

We can rewrite Eq. (3.23) in the following ways:

$$\mathrm{MI_A} = \sum_{w',w,a} p(w', w, a) \log \frac{p(w'|w, a)}{p(w'|w)} \tag{3.24}$$

$$= I(W'; A|W) \tag{3.25}$$

$$= H(W'|W) - H(W'|A, W) \tag{3.26}$$

Equation (3.24) shows that this measure can be computed from observations of W, A, and W' alone. By this we mean, given the sequence of actions over time, $\mathbf{a} = (a_1, a_2, \ldots, a_n)$, and the sequence of world states over time $\mathbf{w} = (w_1, w_2, \ldots, w_n)$, we can sample the joint distribution $p(w', w, a)$. Given the joint distribution, both conditional probability distributions $p(w'|w, a)$ and $p(w'|a)$ can be calculated from the joint distribution in the following way:

$$p(w, a) = \sum_{w'} p(w', w, a) \qquad p(w'|w, a) = \frac{p(w', w, a)}{p(w, a)} \tag{3.27}$$

$$p(w) = \sum_{w',a} p(w', w, a) \qquad p(w'|w) = \sum_{a} \frac{p(w', w, a)}{p(w)}, \tag{3.28}$$

for $p(w), p(w, a) > 0$. Equation (3.25) is shown to point out that $\mathrm{MC_W}$ is the conditional mutual information $I(W'; W|A)$ [59], which is also known as the one-step transfer entropy [18, 60–62].

Equation (3.26) explains what aspects of morphological intelligence are measured by $\mathrm{MI_A}$. The first term is the conditional entropy $H(W'|W)$. If morphological intelligence is high, then this term is minimal, because it refers to the amount of uncertainty about the next world state W' given the current world state W. High uncertainty means that the next world state W' is mostly independent of the current world state W, which contradicts the notion that the body and its interactions with the environment contributed to an observed behaviour. For high morphological intelligence, the next world state W' should be independent of the current action A, which is captured in the second term $H(W'|W, A)$. Hence, to measure morphological intelligence, $\mathrm{MI_A}$ must be inverted. This will be discussed below.

The first important aspect to notice is that the conditional mutual information $I(W'; A|W)$ is upper bounded by the entropy of the next world state $H(W') \leq \log |W'|$ (see Eq. 2.82). This is important because it states that the amount of morphological intelligence, which can be measured by $\mathrm{MI_A}$ is limited by the logarithm of the number of world states. We will use this information to transform $\mathrm{MI_A}$ into a quantification for morphological intelligence. Equation (3.26) shows that $\mathrm{MI_A}$ is zero for maximal morphological intelligence, i.e., when the action A did not have any influence on the next world state (this was discussed above). Hence, to quantify morphological intelligence, $\mathrm{MI_A}$ has to be normalised and inverted in the following way

$$\mathrm{MI'_A} = 1 - \frac{1}{\log |W'|} \mathrm{MI_A}. \tag{3.29}$$

MI'_A is available as part of *gomi* (see Sect. A.2.3).

Up to this point, we implicitly assumed that the system of interest is stochastic. In many applications, e.g. robotics or in particular robot simulations, this is not the case. Therefore, the next step is to investigate how MI_A behaves for deterministic systems. For this purpose, we revisit Eq. (3.26):

$$MI_A = H(W'|W) - H(W'|W, A). \tag{3.30}$$

Given the definition above, it is clear that $H(W'|W)$ and $H(W'|W, A)$ are both equal to zero for deterministic systems, because the conditional entropy $H(X|Y)$ captures the uncertainty about the outcome of the random variable X, if knowledge about Y is available. The following equations show this also analytically for $H(W'|W)$:

$$H(W'|W) = - \sum_{w',w} p(w', w) \log p(w'|w) \tag{3.31}$$

$$= - \sum_{w',w} p(w', w) \log \frac{p(w', w)}{p(w)} \tag{3.32}$$

For deterministic systems, the joint distribution $p(w', w)$ is a Dirac measure, meaning that for any value of w there is only one value of w' for which $p(w', w) > 0$. Hence, we can write:

$$H(W'|W) = - \sum_{w',w} p(w', w) \log \frac{p(w)}{p(w)} = 0 \tag{3.33}$$

Because the conditional mutual information is always larger or equal to zero, i.e., $I(W'; A|W) > 0$ [59], and $I(W'; A|W) = H(W'|W) - H(W'|W, A)$, it follows that $H(W'|W) > H(W'|W, A)$, which means that for deterministic systems $H(W'|W, A) = 0$. To conclude this paragraph, the quantification MI_A will be zero for deterministic systems. This must be kept in mind when MI_A is applied to, e.g. simulated systems without or only very little stochasticity.

In the following sections, we will present a few variations of MI_A. Numerical analysis of the quantifications proposed in this chapter are presented and discussed in the next chapter (see Chap. 4). Applications are discussed in Chap. 5.

3.3.1 Morphological Intelligence as Comparison of Behaviour and Controller Complexity (MI_{MI})

The measure discussed in this section compares the complexity of the behaviour with the complexity of the controller.

The complexity of the behaviour can be measured by the mutual information of consecutive world states, $I(W'; W)$, and the complexity of the controller can be measured by the mutual information of sensor and actuator states, $I(A; S)$, for the following reason. The mutual information of two random variables can also be written as a difference of entropies (see Eq. (2.61)), which, applied to our setting, means that the mutual information $I(W'; W) = H(W') - H(W'|W)$ is high, if we have a high entropy over the next world state W' (first term) but low conditional entropy, i.e., high predictability of W given the current world state W (second term). Summarized, this means that the mutual information $I(W'; W)$ is high if the system shows a diverse but non-random behaviour. The mutual information of a random variable's past and future is also known as predictive information [63, 64]. The one-step predictive information on the sensor data was discussed in the context of diversity-compliance and shown to produce behavioural complexity in embodied agents [47].

Pfeifer and Bongard [10] argued that an embodied system must need to be diverse and compliant in its behaviour to be considered intelligent. The first term ($H(W')$) relates to diversity and the second term ($H(W'|W)$), for the arguments given above, relates to compliance. Hence, a system with high morphological intelligence should produce a complex behaviour based on a controller with low complexity. Formally, this relates to maximising the behavioural complexity $I(W'; W)$ while minimising the policy complexity $I(A; S)$. Minimising $I(A; S)$ either means that the policy has a low diversity in its output (low entropy over actuator states $H(A)$) or that there is only a very low correlation between sensor states S and actuator states A (high conditional entropy $H(A|S)$). Therefore, we define the second measure as the difference between these two terms:

Definition 3.5 (MI$_{MI}$) Let the random variables A, S, W, W' denote the current action, the current sensor state, the current world state and the next world state of an embodied agent, as illustrated in the causal diagram shown in Fig. 3.4. We can then define the contribution of the morphology and its interactions with the environment as the difference between the behavioural complexity and the controller complexity in the following way:

$$\mathrm{MI}_{\mathrm{MI}} = I(W'; W) - I(A; S). \tag{3.34}$$

MI$_{MI}$ is available as part of *gomi* (see Sects. A.2.4 and A.3.3).

Equation (3.34) reveals that this measure is closely related to the work by [33] and in particular his work on relevant information [65], which minimizes the controller complexity while maximising a state-action reward function $Q(s, a)$ in the following way: $\min_{p(a|s)} (I(S; A) - \beta \mathbb{E}\{Q(s, a)\})$. MI$_{MI}$ differs in that the complexity of the behaviour is maximised instead of a reward function.

Note that in the case of a passive observer, i.e., a system that observes the world but in which there is no causal dependency between the action and the next world state (i.e., missing connection between A and W' as shown in Fig. 3.5b), the controller complexity $I(A; S)$ in Eq. (3.34) will reduce the amount of morphological intelligence measured by MI$_{MI}$, although the actuator state does not influence the world dynamics. This might be perceived as a potential shortcoming. In the context discussed of this book, e.g. data recorded from biological or robotic systems, we

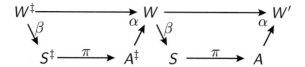

Fig. 3.6 *Two-step causal model of the sensorimotor loop.* Virtual intervention as defined by [53] requires that the parents of the random variable which is the target of the intervention is known. Hence, we expand the single-step sensorimotor loop by the previous time step, denoted by $\ddagger$

think that this will not be an issue. A possible variation of $\mathrm{MI_{MI}}$ that avoids this problem could be $\mathrm{MI'_{MI}} = H(W'|A) - H(W'|W)$. Interestingly, this is the result of a different approach to quantify morphological intelligence within this context, which is discussed next.

3.3.2 MI_{CA}: A Causal Variation of MI_A

In the next step, we apply this formalism of causality to measure the causal information flow between the current world state W, the current action A, and the next world state W'. This requires a two-step causal model of the sensorimotor loop (see Fig. 3.6). The goal is to measure the causal information flow [50] from the current world state W to the next world state W'.

The causal diagram in Fig. 3.6 shows that this causal information flow also passes through the policy, as W and W' are connected via A and S. To quantify morphological intelligence, we want to explicitly make sure not to capture the causal information that flows through the policy, i.e., the system's brain or controller. Hence, we will estimate the causal information flow from the current action A to the next world state W' and subtract it from the causal information that flows from the world state W to the next world state W'. We denote the total causal information flow (including the path over the sensor state S and the current action A) by $CIF(W \rightarrow W')$ and the causal information flow from the current action A to the next world state W' by $CIF(A \rightarrow W')$. The causal variation of $\mathrm{MI_A}$, denoted by $\mathrm{MI_{CA}}$, is then defined in the following way:

$$\mathrm{MI_{CA}} = CIF(W \rightarrow W') - CIF(A \rightarrow W') \tag{3.35}$$

To calculate $CIF(W \rightarrow W')$ and $CIF(A \rightarrow W')$, we need to first derive $p(w'|do(w))$ and $p(w'|do(a))$, which is done in the following paragraph using the formalism by [53] described above.

The starting point is the two-step sensorimotor loop shown in Fig. 3.6. From this, we can derive the joint distribution $p(w', s, a, w, s^{\ddagger}, a^{\ddagger}, w^{\ddagger})$, which is given by

$$p(w', s, a, w, s^{\ddagger}, a^{\ddagger}, w^{\ddagger}) = p(w'|w, a)p(a|s) \tag{3.36}$$
$$p(s|w)p(w|w^{\ddagger}, a^{\ddagger})p(a^{\ddagger}|s^{\ddagger})p(s^{\ddagger}|w^{\ddagger})p(w^{\ddagger})$$

We can now apply the Pearl's formalism:

$$p(w', s, a, w^{\ddagger}, s^{\ddagger}, a^{\ddagger}|do(w)) = p(w'|w, a)p(a|s)p(s|w) \tag{3.37}$$
$$p(a^{\ddagger}|s^{\ddagger})p(s^{\ddagger}|w^{\ddagger})p(w^{\ddagger})$$
$$= p(w'|w, a, s)p(a|s, w)p(s|w) \tag{3.38}$$
$$p(a^{\ddagger}|s^{\ddagger}, w^{\ddagger})p(s^{\ddagger}|w^{\ddagger})p(w^{\ddagger})$$
$$= p(w', a, s|w)p(a^{\ddagger}, s^{\ddagger}, w^{\ddagger}) \tag{3.39}$$

Equation (3.38) results from the fact that W' is independent of S given W and A (see Fig. 3.6), which is formally expressed as $W' \perp\!\!\!\perp S|A, W$. Equation (3.39) can be used to calculate $p(w'|do(w))$ by summing over the variables $W^{\ddagger}$, $S^{\ddagger}$, $A^{\ddagger}$, S, and A:

$$p(w'|do(w)) = \sum_{w^{\ddagger}, s^{\ddagger}, a^{\ddagger}, s, a} p(w^{\ddagger}, w', s^{\ddagger}, a^{\ddagger}, s, a|do(w)) \tag{3.40}$$

$$= \sum_{w^{\ddagger}, s^{\ddagger}, a^{\ddagger}, s, a} p(w', a, s|w)p(a^{\ddagger}, s^{\ddagger}, w^{\ddagger}) \tag{3.41}$$

$$= \sum_{w^{\ddagger}, s^{\ddagger}, a^{\ddagger}} p(a^{\ddagger}, s^{\ddagger}, w^{\ddagger}) \sum_{s, a} p(w', a, s|w) \tag{3.42}$$

$$= p(w'|w) \tag{3.43}$$

The causal information flow can be calculated as the Kullback-Leibler divergence $D(p(w'|do(w))||\hat{p}(w))$ with $p(w'|do(w)) = p(w'|w)$ and the post-interventional distribution $\hat{p}(w') = \sum_w p(w'|do(w))p(w) = \sum_w p(w'|w)p(w) = p(w')$.

The causal information flow from $W \to W'$ is then given by:

$$CIF(W \to W') = \sum_w p(w) \sum_{w'} p(w'|do(w)) \log \frac{p(w'|do(w))}{\hat{p}(w')} \tag{3.44}$$

$$= \sum_w p(w) \sum_{w'} p(w'|w) \log \frac{p(w'|w)}{p(w')} \tag{3.45}$$

$$= I(W'; W) \tag{3.46}$$

Analogous calculations for $p(w'|do(a))$ lead to the following result:

$$p(w^{\ddagger}, w', w, s, a^{\ddagger}, s^{\ddagger}|do(a)) = p(w'|w, a)p(s|w)p(w|w^{\ddagger}, a^{\ddagger}) \qquad (3.47)$$
$$p(a^{\ddagger}|s^{\ddagger})p(s^{\ddagger}|w^{\ddagger})p(w^{\ddagger})$$
$$= p(w'|w, a)p(s|w, w^{\ddagger}, a^{\ddagger})p(w|w^{\ddagger}, a^{\ddagger}, s^{\ddagger}) \quad (3.48)$$
$$p(a^{\ddagger}|s^{\ddagger}, w^{\ddagger})p(s^{\ddagger}|w^{\ddagger})p(w^{\ddagger}) \qquad (3.49)$$
$$= p(w'|w, a)p(s|w, w^{\ddagger}, s^{\ddagger}, a^{\ddagger}) \qquad (3.50)$$
$$\cdot\, p(w, w^{\ddagger}, a^{\ddagger}, s^{\ddagger})$$
$$= p(w'|w, a)p(s, w, w^{\ddagger}, s^{\ddagger}, a^{\ddagger}) \qquad (3.51)$$
$$= p(w'|w, a, s, w^{\ddagger}, s^{\ddagger}, a^{\ddagger}) \qquad (3.52)$$
$$\cdot\, p(s, w, w^{\ddagger}, s^{\ddagger}, a^{\ddagger})$$
$$= p(w', w, s, w^{\ddagger}, s^{\ddagger}, a^{\ddagger}|a) \qquad (3.53)$$
$$p(w'|do(a)) = \sum_{w', w^{\ddagger}, s, a^{\ddagger}, s^{\ddagger}} p(w^{\ddagger}, w', w, s^{\ddagger}, a^{\ddagger}, s|do(a)) \quad (3.54)$$
$$= \sum_{w', w^{\ddagger}, s, a^{\ddagger}, s^{\ddagger}} p(w', w, s, w^{\ddagger}, s^{\ddagger}, a^{\ddagger}|a) \quad (3.55)$$
$$= p(w'|a). \qquad (3.56)$$

The causal information flow from the current action A to the next world state W' can then be quantified in the following way:

$$CIF(A \to W') = \sum_{a} p(a) \sum_{w'} p(w'|a) \log \frac{p(w'|a)}{p(w')} \qquad (3.57)$$
$$= I(W'; A) \qquad (3.58)$$

Finally, we subtract $CIF(A \to W')$ from $CIF(W \to W')$ as given in the definition (see Eq. (3.35)) above, which results in

$$\mathrm{MI_{CA}} = CIF(W \to W') - CIF(A \to W') \qquad (3.59)$$
$$= I(W'; W) - I(W'; A) \qquad (3.60)$$

Hence, the causal variation of $\mathrm{MI_A} = I(W'; A|W)$ is $\mathrm{MI_{CA}} = I(W'; W) - I(W'; A)$, and it is summarised in the following definition:

Definition 3.6 ($\mathrm{MI_{CA}}$) Let the random variables A, W, W' denote the current action, the current world state and the next world state of a reactive embodied agent, as illustrated in the causal diagram shown in Fig. 3.4. We then define the contribution of the morphology and its interactions with the environment as the difference of the causal information flow between the world states and the causal information flow from the current action to the next world state in the following way:

$$\mathrm{MI_{CA}} = CIF(W \to W') - CIF(A \to W') \tag{3.61}$$

$$= I(W'; W) - I(W'; A) \tag{3.62}$$

$\mathrm{MI_{CA}}$ is available as part of *gomi* (see Sects. A.2.5 and A.3.4).

The final two variations discussed in this section are agent-intrinsic formulations of $\mathrm{MI_A}$ and $\mathrm{MI_{CA}}$.

3.3.3 Agent-Intrinsic Variations of MI$_A$ and MI$_{CA}$

The previous quantifications require direct knowledge of the world process, i.e., access to information about the current world state W and the next world state W'. There are contexts in which knowledge about the world states is not available, e.g. robots acting in the real world. In such cases, the agent only has access to information that is provided by the sensors, i.e., which is available by the means of the sensor state S. For this reason, the measures derived in this section will only operate on intrinsically available information, i.e., S, A, and C (see Figs. 3.3 and 3.5).

As the sensor state S is the agent's internal representation of the world, the first variation is given by simply replacing the world state W by the sensor state S and the next world state W' by the next sensor state S' (in the causal diagram shown in Fig. 3.5). The resulting quantification is defined in the following way.

Definition 3.7 (ASOC$_A$) Let the random variables A, S, S' denote the action, the current and the next sensor state of an embodied agent, which is described by the causal diagram shown in Fig. 3.4. The quantification of the morphological intelligence as an associative measure of the negative effect of the action is then defined as:

$$\mathrm{ASOC_A} = D(p(s'|s, a)||p(s'|s)). \tag{3.63}$$

The Kullback-Leibler divergence used in the definition of ASOC$_A$ (see Definition 3.7) is also known as the conditional mutual information $I(S'; A|S)$ and the transfer entropy of A on S [60]. It was investigated in the context of information self-structuring of embodied agents by Lungarella et al. [66].

Following the idea of $\mathrm{MI_{CA}}$ (see Definition 3.6), the second variation is based on measuring the causal information flow of the sensor and action states S, A on the next sensor state S'. To simplify the argumentation, we first consider a reactive system (see Fig. 3.4). The measure will then also be presented and discussed for non-reactive systems, as both cases result in different measures. The idea for this measure is captured in Fig. 3.7. Note, that the sensor states S and S' are here understood as internally available information about the system's environment. The causal information flow from S to S', denoted by $CIF(S \to S')$, includes the information that flows from S to S' over all pathways. This explicitly includes the information flow from S to S' over the action A (see Fig. 3.7a). Morphological intelligence is here

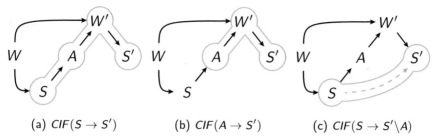

(a) $CIF(S \rightarrow S')$ (b) $CIF(A \rightarrow S')$ (c) $CIF(S \rightarrow S' \backslash A)$

Fig. 3.7 *Visualisation of the causal measure* $\mathrm{MI_{CA}}$. The causal graph used in the Figs. (a), (b), and (c) is the reduction of the sensorimotor loop shown in Fig. 3.4 to two consecutive time steps. Figure **a** shows that the causal information flow $CIF(S \rightarrow S')$ measures all causal information from S to S', including the information that flows over A. Figure **b** shows that the causal information flow $CIF(A \rightarrow S')$ only captures the information flowing from A to S'. Both can be used to approximate the causal information from S to S', that does not pass through A, denoted by $CIF(S \rightarrow S' \backslash A)$, as shown in Figure (**c**)

understood as the causal information flow from S to S' excluding the causal information flowing from the action A to S' (see Fig. 3.7c). Hence, we need to subtract the causal information flow from A to S', denoted by $CIF(A \rightarrow S')$ (see Fig. 3.7b) to exclude it from $CIF(S \rightarrow S')$ such that we receive the causal information flow that goes from S to S' without passing A, denoted by $CIF(S \rightarrow S' \backslash A)$ (see Fig. 3.7c). The derivations are analogous to the derivations required for $\mathrm{MI_{CA}}$ (see Sect. 3.3.2).

According to the theory of causality by Pearl [53], we talk about identifiable causal effects if they are computable from observational data. For the reasons already discussed above, we are interested in deriving a measure that operates on intrinsically available information only. Hence, we require that the causal effects are identifiable from observational data that are intrinsically available to the agent. We refer to this type of identifiability as *intrinsically identifiable*. Next, we show that the causal effects of S on S' and of A on S' are intrinsically identifiable by the following conditional probability distributions:

$$p(s'|\mathrm{do}(a)) = \sum_s p(s'|s, a)p(s) \tag{3.64}$$

$$p(s'|\mathrm{do}(s)) = \sum_a p(a|s)p(s'|\mathrm{do}(a)) \tag{3.65}$$

To derive these results, we apply the same methodology that was used for $\mathrm{MI_{CA}}$ above. We first derive the equation for $p(s'|\mathrm{do}(a))$:

$$p(s'|\text{do}(a)) = \sum_{w,s,w'} p(w,s,w',s'|\text{do}(a)) \tag{3.66}$$

$$= \sum_{w,s,w'} p(w)p(s|w)p(w'|w,a)p(s'|w') \tag{3.67}$$

$$= \sum_{w,s,w'} p(w)p(s|w)p(a|s)p(w'|w,a)p(s'|w')\frac{1}{p(a|s)} \tag{3.68}$$

$$= \sum_{w,s,w'} p(w)p(s|w)p(a|s,w)p(w'|w,a,s)p(s'|w',a,s,w)\frac{1}{p(a|s)} \tag{3.69}$$

$$= \sum_{w,s,w'} \frac{p(w,s,a,w',s')}{p(s,a)}p(s) \tag{3.70}$$

$$= \sum_{s} p(s'|s,a)p(s) \tag{3.71}$$

Next, we present the calculations for $p(s'|\text{do}(s))$:

$$p(s'|\text{do}(s)) = \sum_{w,a,w'} p(w,a,w',s'|\text{do}(s)) \tag{3.72}$$

$$= \sum_{a} p(a|s) \sum_{w,w'} p(w)p(w'|w,a)p(s'|w') \tag{3.73}$$

$$= \sum_{a} p(a|s) \sum_{w,w'} \left(\sum_{s^{\ddagger}} p(s^{\ddagger})p(w|s^{\ddagger}) \right) p(w'|w,a)p(s'|w') \tag{3.74}$$

$$= \sum_{a} p(a|s) \sum_{s^{\ddagger}} p(s^{\ddagger}) \sum_{w,w'} p(w|s^{\ddagger})p(w'|w,a)p(s'|w') \tag{3.75}$$

$$= \sum_{a} p(a|s) \sum_{s^{\ddagger}} p(s^{\ddagger}) \sum_{w,w'} p(w|s^{\ddagger},a)p(w'|w,a,s^{\ddagger})p(s'|w') \tag{3.76}$$

$$= \sum_{a} p(a|s) \sum_{s^{\ddagger}} p(s^{\ddagger})p(s'|s^{\ddagger},a) \tag{3.77}$$

$$= \sum_{a} p(a|s)p(s'|\text{do}(a)) \tag{3.78}$$

It is important to note, that the equations above (see Eqs. (3.64) and (3.65)) allow us to determine the causal effects *without* any actual intervention. This means that an agent can act in the sensorimotor loop, and from its observation determine e.g. the causal effect of its actions A on its next sensor states S'. From the two probability distributions given in Eqs. (3.64) and (3.65) we can construct the two causal information measures for $CIF(S \to S')$ and $CIF(A \to S')$ (see also [17]). The causal information flow from the current sensor state S to the next sensor state S' ($CIF(S \to S')$) and the current action A to the next sensor state S' ($CIF(A \to S')$)

can now be formalised in the following way:

$$CIF(S \rightarrow S') = \sum_s p(s) \sum_{s'} p(s'|do(s)) \log \frac{p(s'|do(s))}{\sum_{s^\ddagger} p(s'|do(s^\ddagger)) p(s^\ddagger)} \quad (3.79)$$

$$CIF(A \rightarrow S') = \sum_a p(a) \sum_{s'} p(s'|do(a)) \log \frac{p(s'|do(a))}{\sum_{a'} p(s'|do(a')) p(a')} \quad (3.80)$$

The difference $CIF(S \rightarrow S') - CIF(A \rightarrow S')$ is always negative, as the following calculations show:

$$CIF(S \rightarrow S') = \sum_{s'} p(s'|do(s)) \log \frac{p(s'|do(s))}{\sum_{s'} p(s^\ddagger) p(s'|do(s'))} \quad (3.81)$$

$$= \sum_{s'} \left(\sum_a p(a|s) p(s'|do(s)) \right) \quad (3.82)$$

$$\cdot \log \frac{\sum_a p(a|s) p(s'|do(a))}{\sum_{s^\ddagger} p(s^\ddagger) \sum_a p(a|s) p(s'|do(a))} \quad (3.83)$$

$$= \sum_{s'} \left(\sum_a p(a|s) p(s'|do(a)) \right) \log \frac{\sum_a p(a|s) p(s'|do(a))}{\sum_a p(a) p(s'|do(a))} \quad (3.84)$$

$$\leq \sum_a p(a|s) \sum_{s'} p(s'|do(a)) \log \frac{p(s'|do(a))}{\sum_a p(a) p(s'|do(a))} \quad (3.85)$$

$$\Rightarrow CIF(S \rightarrow S') \leq \sum_a p(a|s) CIF(a \rightarrow S'). \quad (3.86)$$

Hence, the resulting measure must be inverted, which leads to the following definition.

Definition 3.8 (C_A) Let the random variables A, S, S' denote the action, the current and the next sensor state of a reactive embodied agent, which is described by the causal diagram shown in Fig. 3.4. The quantification of the morphological intelligence as a causal measure of the negative effect of the action is then defined as:

$$C_A := 1 + \frac{1}{\log |\mathcal{S}|} (CIF(S \rightarrow S') - CIF(A \rightarrow S')) \quad (3.87)$$

$$= 1 - \frac{1}{\log |\mathcal{S}|} D(p(s'|do(a)) || p(s'|do(s))) \quad (3.88)$$

$$= 1 - \frac{1}{\log |\mathcal{S}|} \sum_{s,a} p(s, a) \sum_{s'} p(s'|do(a)) \log \frac{p(s'|do(a))}{p(s'|do(s))} \quad (3.89)$$

C_A is available as part of *gomi* (see Sect. A.2.6).

To date, it is not clear, if the causal information flow $CIF(S \rightarrow S')$ is intrinsically identifiable for non-reactive systems [37]. Therefore, we consider the causal information flow from the internal controller state C to the next sensor state S' (see Fig. 3.3). This is valid, because C represents the entire history of the system, and therefore, also the internal representation of the entire history of the world. All further calculations are analogous to the previous case by replacing S with C (see also [37]) and lead to the following definition.

Definition 3.9 (C_A^d) Let the random variables A, C, S' denote the action, the controller and the next sensor state of a non-reactive or deliberative embodied agent, which is described by the causal diagram shown in Fig. 3.3. The quantification of the morphological intelligence as a causal measure of the negative effect of the action is then defined as:

$$C_A^d := 1 + \frac{1}{\log |\mathcal{S}|}(CIF(C \rightarrow S') - CIF(A \rightarrow S')) \tag{3.90}$$

$$= 1 - \frac{1}{\log |\mathcal{S}|}D(p(s'|\mathrm{do}(a))\|p(s'|\mathrm{do}(c))) \tag{3.91}$$

$$= 1 - \frac{1}{\log |\mathcal{S}|}\sum_{c,a} p(c,a) \sum_{s'} p(s'|\mathrm{do}(a)) \log \frac{p(s'|\mathrm{do}(a))}{p(s'|\mathrm{do}(c))} \tag{3.92}$$

The last definition (see Definition 3.9) is given for completeness only. Morphological intelligence is mainly discussed in the context of behaviours which are well-modelled as reactive behaviours (e.g. locomotion). To the best of our knowledge, it has so far not been discussed in the context of non-reactive or deliberative behaviours.

This concludes the measures of the first concept, in which morphological intelligence is calculated inversely proportional to the influence of the current action A on the next world state W'.

3.4 Concept Two: Quantifying Morphological Intelligence as the Contribution of the World to Itself

Trying to isolate the strength of the causal dependence between the current world state W and the next world state W' (see Fig. 3.3) leads to the formalisation of the second concept. By strength, we mean how much the current world state W influences the next world state W'. In case of the Passive Dynamic Walker, discussed in the previous chapter (see Chap. 1), it is obvious to see that the current world state W fully determines the next world state W' as there is no action A that contributes to the observed behaviour. To quantify the strength of the causal link between W and W' we measure the difference of the observation from the assumption that the link is not present. This is illustrated in Fig. 3.8 and explained in the following paragraphs.

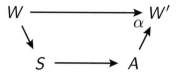
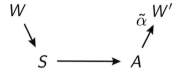

(a) Full Model: This causal diagram is the one-step version of the sensorimotor loop for reactive systems shown in Figure 3.4.

(b) Model in which the world state W has no influence on the next world states W'

Fig. 3.8 *Visualisation of MI_W.* This figure shows the two models of the sensorimotor loop that are compared in the measure

As in the concept discussed in the previous section, the starting point for the considerations is the world dynamics kernel $\alpha(w'|w, a)$, which captures the influence of the actuator signal A and the previous world state W on the next world state W' (see Sect. 3.2). A complete absence of morphological intelligence would mean that the behaviour of the system is entirely determined by the system's controller, and hence, by the actuator state A. In this case, the world dynamics kernel reduces to $\tilde{\alpha}(w'|a)$ (see Fig. 3.8), which can be calculated in the following way:

$$\tilde{\alpha}(w'|a) = \sum_w \alpha(w'|w, a) p(w|a). \qquad (3.93)$$

Any divergence from this assumption means that the previous world state W had an influence, and hence, information about W changes the distribution over the next world states W'. The discrepancy of these two distributions can be measured with the Kullback-Leibler divergence $D_{KL}(\alpha(w'|w, a)||p(w'|a))$. This leads to the following definition.

Definition 3.10 (MI_W) Let the random variables A, W, W' denote the action, the current and the next world state of an embodied agent, which is described by the causal diagram shown in Fig. 3.8a. The quantification of the morphological intelligence as the positive effect of the world onto itself is defined as

$$MI_W := D(p(w'|w, a)||p(w'|w)) \qquad (3.94)$$

MI_W is available as part of *gomi* (see Sects. A.2.1 and A.3.1).

Equation (3.94) can be rewritten in the following way

$$\text{MI}_W = I(W'; W|A) \tag{3.95}$$

$$= \sum_{w',w,a} p(w', w, a) \log \frac{p(w'|w, a)}{p(w'|a)} \tag{3.96}$$

$$= \underbrace{H(W'|A)}_{\text{Independence}} - \underbrace{H(W'|A, W)}_{\text{Predictability}} \tag{3.97}$$

Equation (3.95) restates that MI_W is the conditional mutual information of the world states given the action state $I(W'; W|A)$.

Equation (3.96) shows that this measure can be computed from observations of W, A, and W' alone. By this, we mean that given the sequence of actions over time, $\mathbf{a} = (a_1, a_2, \ldots, a_n)$, and the sequence of world states over time $\mathbf{w} = (w_1, w_2, \ldots, w_n)$, we can sample the joint distribution $p(w', w, a)$. Both conditional probability distributions $p(w'|w, a)$ and $p(w'|a)$ can be calculated from the joint distribution in the following way

$$p(w, a) = \sum_{w'} p(w', w, a) \qquad p(w'|w, a) = \frac{p(w', w, a)}{p(w, a)} \tag{3.98}$$

$$p(a) = \sum_{w',w} p(w', w, a) \qquad p(w'|a) = \sum_{w} \frac{p(w', w, a)}{p(a)}, \tag{3.99}$$

for $p(a), p(w, a) > 0$.

Equation (3.97) is interesting because it highlights how morphological intelligence is understood in the context of this section. The first term, which is labelled *independence*, is the conditional entropy $H(W'|A)$.

High morphological intelligence is only possible, with respect to MI_W, if the conditional entropy $H(W'|A)$ is large. This means that the uncertainty about the next state W' must be high if all we have is information about the current action A. In other words, the next world state W' must be independent of the current action A. The problem here is that a high conditional entropy $H(W'|A)$ could be achieved by completely decoupling the controller from the embodiment. Generally, this is not a desired behaviour, because it would mean that the behaviour of the system is completely independent of the agent's policy. The second term $H(W'|W, A)$ prevents this for the following reason.

To maximise MI_W the first term must be maximised, and the second term, labelled *predictability*, must be minimised. Minimising the conditional entropy $H(W'|W, A)$ means that the surprise or uncertainty about the next world state W' should be zero if the current world state W and the current action A are known. In other words, the next world state W' should be fully predictable, given the current world state W and the current action A. We will further investigate MI_W below when we discuss morphological intelligence in the context of information decomposition (see Sects. 3.4.1 and 3.5). Next, we will investigate the upper bound and the behaviour of the quantification for deterministic systems analogous to our calculations for MI_A (see previous Sect. 3.3).

The conditional mutual information $I(W'; W|A)$ is bounded by the entropy of the next world state $H(W') \leq \log|W'|$ (see Theorem. 2.82). This is important because it states the amount of morphological intelligence that can be measured by MI_W is limited by the logarithm of the number of world states.

Up to this point, we implicitly assumed that our systems are stochastic. As stated earlier, in many applications, e.g. robotics or in particular robot simulations, this is not the case. Therefore, the next step is to investigate how MI_W behaves for deterministic systems. For this, we revisit Eq. (3.97):

$$MI_W = H(W'|A) - H(W'|W, A). \tag{3.100}$$

For fully deterministic systems, the maps $\beta(s|w)$ and $\pi(a|s)$ are Dirac measures, meaning that for any value of w there is only one value of a for which $p(w, a) > 0$. Hence, for deterministic systems, the following equality holds

$$H(W'|W, A) = H(W'|A). \tag{3.101}$$

It then follows from Eqs. (3.100) and (3.101) that

$$MI_W = H(W'|A) - H(W'|W, A) \tag{3.102}$$
$$= H(W'|A) - H(W'|A) \tag{3.103}$$
$$= 0. \tag{3.104}$$

The conclusion is that MI_W is zero for fully deterministic systems. This should be taken into account if the measure is applied to systems without or only very limited stochasticity.

In the next section, we will investigate MI_W in the context of information decomposition as it was introduced by [67].

3.4.1 Information Decomposition of MI_W

This section describes the decomposition of multivariate mutual information as defined by [68, 69]. It was first applied to the sensorimotor loop by Ghazi-Zahedi and Rauh [70]. A different form of information decomposition based on the work by Ay [71] is discussed in Sect. 3.5.

Consider three random variables X, Y, Z. Suppose that the goal is to predict the value of the random variable X, based on the information that is available through knowledge of Y and Z. The question is, how is the information that Y and Z carry about X distributed over Y and Z? In general, there may be *redundant* or *shared* information (information contained equally in both Y and Z), but there may also be *unique* information (information contained in either Y or Z). Finally, there is also the possibility of *synergistic* or *complementary* information, i.e. information that is only available when Y and Z are taken together. The classical example for synergistic information is the XOR function: If Y and Z are binary random variables and if $X = Y \text{ XOR } Z$, then neither Y nor Z contain any information about X (in fact, X is independent of Y and X is independent of Z), but when Y and Z are taken together, they completely determine X.

The total information that (Y, Z) contains about X can be quantified by the mutual information $I(X; Y, Z)$. However, there is no canonical way to separate the four kinds of information described above. Different variations have been proposed (see e.g. [67–69, 71–74]), but a final definition has yet to be found.

Mathematically, one would like to have four functions, namely shared information $(SI(X : Y; Z))$, unique information of Y $(UI(X : Y \setminus Z))$, unique information of Z $(UI(X : Z \setminus Y))$, and finally, synergistic or complementary information that satisfy

$$I(X : (Y, Z)) = SI(X : Y; Z) + UI(X : Y \setminus Z) \quad (3.105)$$
$$+ UI(X : Z \setminus Y) + CI(X : Y; Z).$$

The decomposition shown in Eq. (3.105) was first described in [67]. It follows from Eq. (3.105) and the chain rule of mutual information that an information decomposition always satisfies

$$I(X : Y|Z) = UI(X : Y \setminus Z) + CI(X : Y; Z). \quad (3.106)$$

Several candidates have been proposed for SI, UI, and CI so far (see e.g. [67, 73]). A different candidate will be presented below (see Sect. 3.5.2). In this section, we will describe the decomposition of [68], which is defined in the following way.

Let Σ be the set of all possible joint distributions of X, Y, and Z. Fix an element $P \in \Sigma$ (the "true" joint distribution of X, Y, and Z).

Define

$$\Lambda_P = \left\{ Q \in \Sigma : Q(X = x, Y = y) = P(X = x, Y = y) \right.$$

$$\text{and } Q(X = x, Z = z) = P(X = x, Z = z)$$

$$\left. \text{for all } x \in \mathcal{X}, y \in \mathcal{Y}, z \in \mathcal{Z} \right\} \quad (3.107)$$

as the set of all joint distributions which have the same marginal distributions on the pairs (X, Y) and (X, Z). Then

$$UI(X : Y \setminus Z) = \min_{Q \in \Lambda_P} I_Q(X : Y | Z), \tag{3.108}$$

$$SI(X : Y; Z) = \max_{Q \in \Lambda_P} CoI_Q(X; Y; Z), \tag{3.109}$$

$$CI(X : Y; Z) = I(X : Y, Z) - \min_{Q \in \Lambda_P} I_Q(X : (Y, Z)), \tag{3.110}$$

where CoI denotes the co-information as defined in [75]. Here, a subscript Q in an information quantity means that the quantity is computed with respect to Q as the joint distribution.

In [68], the equations for UI, CI, and SI are derived from considerations about decision problems in which the objective is to predict the outcome of X. To apply this to the decomposition of $\mathrm{MI_W}$, we will set $X = W'$, $Y = W$, and $Z = A$. In the context of the sensorimotor loop, W and A not only have information about W', but they actually *control* W'. However, from an abstract point of view, the situation is similar. In the sensorimotor loop, we also expect to find aspects of redundant, unique, and complementary influence of W and A on W'. Formally, since everything is defined probabilistically, we can still use the same functions UI, CI, and SI. We believe that the arguments behind the definition of UI, CI and SI remain valid in the setting of the sensorimotor loop where we need it.

We can now give two definitions of quantifying morphological intelligence that refer to the synergistic and unique information:

Definition 3.11 $(UI(W' : W \setminus A))$ Let the random variables A, W, W' denote the current action, the current and the next world state of an embodied agent, which is described by the causal diagram shown in Fig. 3.4. The quantification of the unique information that the current world state W has about the next world state W', which is not present in the current action A is then defined as:

$$UI(W' : W \setminus A) = \min_{Q \in \Lambda_P} I_Q(W' : W | A), \tag{3.111}$$

where Q is defined according to Eq. (3.107).

Quantifying morphological intelligence as the synergistic information body and brain is defined in the following way:

Definition 3.12 $(CI(W' : W; A))$ Let the random variables A, W, W' denote the current action, the current and the next world state of an embodied agent, which is described by the causal diagram shown in Fig. 3.4. The quantification of morphological intelligence as the synergy between body and brain (with respect to [68]) is then defined as:

$$CI(W' : W; A) = I(W' : W, A) - \min_{Q \in \Lambda_P} I_Q(W' : (W, A)), \tag{3.112}$$

where Q is defined according to Eq. (3.107).

The reason for investigating the unique and synergistic information is indicated in Eq. (3.106), which we will rewrite here in relation to the sensorimotor loop in the following way:

$$\text{MI}_\text{W} = I(W'; W|A) \tag{3.113}$$
$$= UI(W' : W\backslash A) + CI(W' : W; A) \tag{3.114}$$

Equation (3.114) shows that MI_W can be decomposed into unique and synergistic information. In this section, we discuss the unique information $UI(W' : W\backslash A)$, which is the information that the current world state W contains about the next world state W', that is not available through knowledge about the current action A. The unique information is only available if the state of the current world state W is known.

The most often cited definition for morphological computation describes processes which are conducted by the body that otherwise would have to be conducted by the brain [10] (see Chap. 1). In this sense, morphological computation can be understood as information processing that occurs only in the body and contributes to behaviour. This aspect of morphological computation, and hence also morphological intelligence, is captured by the unique information $UI(W'; W\backslash A)$. It quantifies how a process that can only be attributed to the morphology and its interaction with the environment, contributed to the observed behaviour. A detailed motivation to quantify morphological intelligence as the synergy between body and brain is given in the next section, which is based on the formalism introduced in [71]. The reason it is discussed in the next section will be discussed below.

Equation (3.114) also shows the advantage of the information decomposition. Synergistic and unique information can be computed from each other if the conditional mutual information $I(W'; W|A)$ is known. The conditional mutual information can be easily derived from observation (see Sect. 3.3 and Eq. (3.24)), given that there are enough samples with respect to the dimensionality of W', W, A. We are not only interested in mathematical rigorous definitions, but we are also interested applicability to real data. And this is where the decomposition by Bertschinger et al. [68] currently has a disadvantage. There is no algorithm known to us to compute synergistic and unique information for non-trivial, i.e., non-binary systems. Proposals to calculate synergistic and unique information which build upon standard optimisation techniques have been proposed in [76]. For the binary model of the sensorimotor loop (see below), we used an approximation in our previous work [70] that was already described in the original paper [68].

This concludes the analysis of MI_W in terms of an information decomposition. Next, we ask how an agent can quantify morphological intelligence as the contribution of the world onto itself from its intrinsic perspective, i.e., without access to the world states W and W'. This can be useful, e.g. in the context of learning, when the agent has to maximise morphological intelligence based on intrinsically available information only. Other examples are real robots as they can only access information about the environment through their sensors, i.e., they only have access to W through their sensor state S.

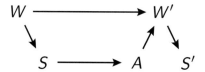

Fig. 3.9 *Extension of the one-step sensorimotor loop for ASOC$_W$. This figure shows the causal diagram of the sensorimotor loop that is required for ASOC$_W$. The one-step sensorimotor loop is extended by the next sensor state S' because ASOC$_W$ operates only on information that is intrinsically available to the agent, and hence, requires S and S' as substitutions for W and W'*

3.4.2 Agent-Intrinsic Variation of MI$_W$

In the context of this book, agent-intrinsic means that the measures may only operate on information that is intrinsically available to the agent. Given the sensorimotor loop described above (see Fig. 3.3), this means that the measures discussed in this section can only access on the random variables S, A, and C.

The first measure is a translation of MI$_W$ to the intrinsic perspective by substituting the current world state W by the current sensor state S. For this purpose, we have to extend the sensorimotor that is shown in Fig. 3.8a to include the next sensor state S' (see Fig. 3.9). We can now define the first variation of MI$_W$ that only uses information that is intrinsically available to the agent, which is called ASOC$_W$.

Definition 3.13 (ASOC$_W$) Let the random variables A, S, S' denote the action, the current and the next sensor state of an embodied agent, which is described by the causal diagram shown in Fig. 3.9. The quantification of morphological intelligence as an associative measure of the positive effect of the world on itself is then defined as:

$$\text{ASOC}_W := D(p(s'|s,a)\|p(s'|a)) \tag{3.115}$$

$$= \sum_{s',s,a} p(s',s,a) \log \frac{p(s'|s,a)}{p(s'|a)} \tag{3.116}$$

$$= I(S'; S|A) \tag{3.117}$$

A non-trivial variation of MI$_W$ to the agent's intrinsic perspective is discussed next. It requires a world model $p(s'|s,a)$, which is the estimation of the next sensor state S', given the current sensor state S and the current action A. In other words, the internal world model is an internal representation of the world dynamics kernel $\alpha(w'|w,a)$. In general, the internal world model $p(s'|s,a)$ is insufficient, because the sensor states only capture a portion of the information that is available in the world state. Yet, is it the only possible way, that the agent can make predictions about the future without full access to the world states. Figure 3.10 visualises the idea that leads to the next variation of MI$_W$, denoted by C$_W$, which only uses intrinsically available information. The internal world model $p(s'|s,a)$ allows the agent to estimate the

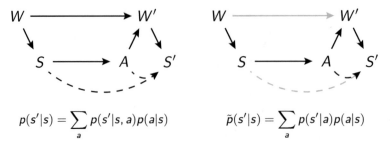

Fig. 3.10 *Visualisation of the conditional independence measure* C_W. The left-hand side shows how the conditional probability distributions $p(s'|s)$ can be calculated from the world model $p(s'|s, a)$ and the policy $p(a|s)$. The right-hand side shows how $p(s'|s)$ changes, if one assumes that the world does not influence itself (grey arrow between W and W') and if this is reflected in the internal world model (grey arrow between S and S'). The difference between the two models can be used to quantify morphological intelligence

conditional probability distribution $p(s'|s)$, which is a proxy for the behaviour of the system. The conditional probability distribution $p(s'|s)$ is given by:

$$p(s'|s) = \sum_a p(s'|s, a)p(a|s). \qquad (3.118)$$

Analogous to MI_W, we now assume that the causal link between the current world state W and the next world state W' is not present, i.e., that there is no morphological intelligence present in the observed behaviour. This means that the intrinsic world model no longer depends on the current sensor state S. The resulting conditional probability distribution, denoted by $\tilde{p}(s'|s)$ is given by:

$$\tilde{p}(s'|s) = \sum_a p(s'|a)p(a|s) \qquad (3.119)$$

$$= \sum_a p(a|s) \sum_{s^{\ddagger}} p(s'|s^{\ddagger}, a)p(a|s^{\ddagger})\frac{p(s^{\ddagger})}{p(a)} \qquad (3.120)$$

Analogous to MI_W, C_W is then defined as the Kullback-Leibler divergence of $p(s'|s)$ and $\tilde{p}(s'|s)$.

Definition 3.14 (C_W) Let the random variables A, S, S' denote the action, the current and the next sensor state of an embodied agent, which is described by the causal diagram shown in Fig. 3.10. The quantification of morphological intelligence as the divergence from the assumption, that the next world state is conditionally independent of the previous world state is then defined as:

$$C_W := D(p(s'|s)||\hat{p}(s'|s)) \tag{3.121}$$

$$= \sum_{s',s} p(s'|s)p(s) \log \frac{p(s'|s)}{\tilde{p}(s'|s)}, \tag{3.122}$$

where $p(s'|s)$ is defined in Eq. (3.118) and $\tilde{p}(s'|s)$ is defined in Eq. (3.120).

This concludes the presentation of the second concept for quantifying morphological intelligence, which measures morphological intelligence as the positive effect of the world on itself.

3.5 Concept Three: Quantifying Morphological Intelligence as Synergy of Body and Brain

This section will give a motivation for the quantification of morphological intelligence as the synergy of body and brain. Formally, this means quantifying morphological intelligence as the synergistic information that the current action A and the current world state W contain about the next world state W'. The relation to the previous concepts, namely MI_W and MI_A, was discussed in the previous section (see Sect. 3.4.1).

The motivation to investigate morphological intelligence as synergistic information rose from the motivation to distinguish between a purely physical system, such as the ball rolling downhill and the Passive Dynamic Walker (see Chap. 1). Both cases are purely mechanical systems, but one would assign morphological computation to the Passive Walker, whereas one would generally have difficulties to state that the ball is performing computation or reducing the computational complexity for a brain.

There are three possible solutions to this problem. Their relation to synergy will be addressed below. First, as argued in [17, 20], the Passive Walker itself is not performing computation, but it shows that morphological computation can be present in human walking.

Second, the definition of physical computation ([28], and Sect. 1.2.3) offers a possibility to distinguish between pure physics and physical computation. With respect to the ball and the Passive Walker, the theory of [28] would lead to the conclusion that both are computing if there is a user (e.g. brain) that translates their states, e.g. to measure the slope.

The third possibility can be summarised in the following way (cited from [77]): "*nonneural body parts could be described as parts of a computational system, but they do not realise computation autonomously, only in connection with some kind of [...] central control system.*" If we now compare the three different approaches, they don't seem to be entirely different. All three cases argue that morphological intelligence requires the close interaction of a brain and body.

Given this motivation, this section will discuss the quantification of morphological intelligence as the amount of synergistic information and will draw connections to the two previous sections. In the previous section, we already looked at an information decomposition of the sensorimotor loop and found that the first two concepts can be written as the sum of unique and synergistic information, i.e.,

$$\text{MI}_W = I(W'; W|A) \tag{3.123}$$

$$= UI(W' : W \setminus A) + CI(W' : W; A) \tag{3.124}$$

$$\text{MI}_A = I(W'; A|W) \tag{3.125}$$

$$= UI(W' : A \setminus W) + CI(W' : W; A) \tag{3.126}$$

where $UI(W' : W \setminus A)$ is the unique information that the current world state W carries about the next world state W' that cannot be gathered through knowledge about the current action A, and $UI(W' : W \setminus A)$ is analogously the information about the next world state W' that is only available through knowledge about the current action A, and finally $CI(W' : W; A)$ is the information about the next world state W' that is only available if both, the current action A and the current world state W are known.

There are two problems with the information decomposition as it was proposed by [68]. First, applicability to non-binary systems is still an unsolved problem. This is important because we are not only interested in a theoretical foundation but we are also interested in applicability to real-world systems. The second problem is that in their definition, the authors do not incorporate the probability distributions over the inputs, in this case, the random variables W and A, which means that the synergistic measure $CI(W' : W; A)$ can falsely detect correlations in the input as synergistic information (this will be discussed in more detail below and shown in the next chapter, see Sect. 4.2).

These two disadvantages are the reason we discuss a measure for synergistic information in the sensorimotor loop based on the complexity measure defined by Ay [71], which is discussed next. The basic idea of the complexity measure is summarised in the paper in the following way: "*The whole is more than the sum of its elementary parts*" [71]. The underlying information theoretic idea is best explained along with the schematics shown in Fig. 3.11.

The left-hand side of Fig. 3.11 shows the "*whole*", while the right-hand side shows "*the elementary parts*" of a stochastic system with two input variables, X_1 and X_2, and two output variables, Y_1 and Y_2. We refer to the graphical model on the left-hand side of Fig. 3.11 as the *full model*, whereas the model on the right-hand side will be referred to as the *split model*.

The full model assumes that every output node is connected to every input node, whereas the split model assumes that each input node only affects one output node. The undirected connection between the input nodes indicates that the input distribution is taken into account in both models. Both models are defined by their feature sets F. In the example given above, the feature set for the full model is given by $F_{\text{full}} = \{\{X_1, Y_1, X_2, Y_2\}\}$, and the feature set for the split model is given

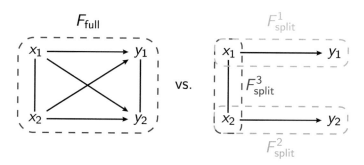

Fig. 3.11 *Quantifying complexity.* Left-hand side: Full model of two input and two output variables. Right-hand side: Split model, as proposed by [71]

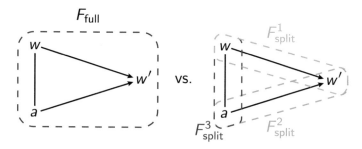

Fig. 3.12 *Quantifying synergy in the sensorimotor loop.* Left-hand side: Full model of two input and two output variables. Right-hand side: Split model, as proposed by [72]

by $F_{\text{split}} = \{\{X_1, Y_1\}, \{X_2, Y_2\}, \{X_1, X_2\}\}$. Note that we have explicitly included the feature corresponding to the input distribution $\{X_1, X_2\}$. The divergence between the full and split models is defined as a measure for complexity by Ay [71]. Variations of this measure have been proposed in [78, 79] and compared in [80].

As discussed throughout this work, we are primarily interested in the relation of the three random variables W, A, and W', which represent the current world state, the current action, and the next world state. Hence, we translate the quantification for complexity [71] which was originally formulated for four random variables (see Fig. 3.11) to three random variables (see Fig. 3.12). The resulting quantification is also known as synergistic information and was first discussed in [72].

Given three random variables X, Y, and Z (see Fig. 3.12), the synergistic information is defined as the Kullback-Leibler divergence

$$\text{MI}_{\text{SY}} = D(p_{\text{full}}(x|y, z) || p_{\text{split}}(x|y, z)), \tag{3.127}$$

where the full and split models are defined by the feature sets depicted in Fig. 3.12.

In their work, the authors showed that the main difference between the synergistic measure defined by Bertschinger et al. [68] and the synergistic measure defined by

Ay [72] is that the former does not take the input distribution into account. This is shown by the following Eq. (cited from [72]):

$$CI(X : Y; Z) = I(X, Y : Z) - \min_{q \in \Delta_P} I_q(X, Y : Z) \qquad (3.128)$$

$$\Delta_P = \Big\{ Q \in \Sigma : Q(X = x, Y = y) = P(X = x, Y = y)$$

$$\text{and } Q(X = x, Z = z) = P(X = x, Z = z)$$

$$\text{and } Q(Y = y, Z = z) = P(Y = y, Z = z)$$

$$\text{for all } x \in \mathcal{X}, y \in \mathcal{Y}, z \in \mathcal{Z} \Big\} \qquad (3.129)$$

The difference between the measure proposed by Bertschinger et al. [68] and Perrone and Ay [72] is seen in line 3 of Eq. (3.129) (compare with Eq. (3.107)).

Applying the measure by Perrone and Ay [72] to the sensorimotor loop and in the context of morphological intelligence, we are asking how much the observation of our embodied agent's behaviour differs from the assumption that there is no synergy between body and brain. This leads us to the following definition.

Definition 3.15 (MI_{SY}) Let the random variables A, W, W' denote the current action, the current world state and the next world state of an embodied agent, as illustrated in the causal diagram shown in Fig. 3.4. Morphological intelligence as the synergy between body and brain is then defined as:

$$\text{MI}_{\text{SY}} = D(p_{\text{full}}(w'|w, a) \| p_{\text{split}}(w'|w, a)), \qquad (3.130)$$

where $p_{\text{full}}(w'|w, a)$ is the conditional probability distribution based on the feature set $F_{\text{full}} = ((W', W, A))$ and $p_{\text{split}}(w'|w, a)$ is the conditional probability distribution based on the feature set $F_{\text{split}} = ((W', W), (W', A), (W, A))$ (see Fig. 3.12).

MI_{SY} is available as part of *gomi* (see Sect. A.2.7).

Next, we discuss how the split model can be calculated. Methods in this context are known as maximum entropy estimators.

3.5.1 Maximum Entropy Estimation with the Iterative Scaling Algorithm

There is a method to calculate the maximum entropy estimation of a probability distribution based on feature sets, known as iterative scaling, which is well-established and goes back to the work of Darroch and Ratcliff [81] and Csiszár [82]. The algorithm can be summarised in the following form (for joint distributions).

Let $\hat{p}$ be the target distribution, which in this case is given by $\hat{p}(w', w, a) = \sum_s p(w)\beta(s|w)\pi(a|s)\alpha(w'|w, a)$, V be the set of random variables (e.g.

$V = \{W', W, A\}$) and F be the feature set (e.g. $F_{split} = \{\{W', W\}, \{W', A\}, \{W, A\}\}$).
For the sake of simplicity in the presentation, we use the following abbreviations:
$p(V) = p(w', w, a)$, $p(F_i)$ is either $p(w, a)$, $p(w', a)$, or $p(w', w)$ (depending on the
index i), and $p(V \backslash F_i | F_i)$ is either $p(w' | w, a)$, $p(w | w', a)$, or $p(a | w', w)$ depending
on the selected feature.

The target distribution (which is approximated with the maximum entropy
method) is denoted by $\hat{p}(V)$. As an example, the target distribution for the first
feature is given by $\hat{p}(F_1) = \sum_a \hat{p}(w', w, a)$. Iterative scaling is then defined in the
following way:

$$p^{(0)}(V) = \frac{1}{|V|} \tag{3.131}$$

$$p^{(n+1)}(V) = \hat{p}(F_{n \bmod |F|}) \cdot p^{(n)}(V \backslash F_{n \bmod |F|} | F_{n \bmod |F|}). \tag{3.132}$$

In other words, we initialise the joint distribution $p^{(0)}$ to be the uniform distribution.
In each iteration, we pick one feature from the set of features. We then multiply the
marginal distribution of this feature in the target distribution ($\hat{p}(F_{n \bmod |F|})$) with the
conditional distribution of the remaining variables conditioned on the chosen feature
($p^{(n)}(V \backslash F_{n \bmod |F|} | F_{n \bmod |F|})$). In the notion above, $F_{n \bmod |F|}$ refers to the iterative
selection of features, based on the current iteration step n, modulo the number $|F|$
of defined features. This algorithm is proved to converge [81, 82], and it is used in
the numerical simulations below (source code is available at [83] and [84]).

This concludes the discussion about the synergistic measure and how it can be
computed. It was stated above, that the unique information can be calculated from the
conditional mutual information $I(W'; W | A)$ and the synergistic information MI_{SY}.
Hence, analogous to the previous section, this section concludes with a discussion
about unique information in the context of morphological intelligence.

3.5.2 Quantifying Unique Information

Equation (3.114) shows that the conditional mutual information $I(X; Y | Z)$ is given
by the sum of the unique information $UI(W' : W \backslash A)$ and the synergistic information $CI(W' : W; A)$. Given that MI_{SY} is a measure for synergy, we can now also
give a new definition for the unique information $UI(W' : W \backslash A)$ in terms of MI_W
and MI_{SY}. We denote this measure by MI_W^P because it captures the part of MI_W that
results from physics, i.e., uncontrolled body-environment interactions. This leads us
to the following definition:

Definition 3.16 (MI_W^P) Let the random variables A, W, W' denote the current action,
the current world state and the next world state of an embodied agent, as illustrated in
the causal diagram shown in Fig. 3.4. The unique information of the world process,
i.e., the unique information that the current world state W carries about the next

world state W' that cannot be extracted from knowledge about any of the other random variables in the causal diagram of the sensorimotor loop, is then given by the following equation

$$\mathrm{MI}_W^p = \mathrm{MI}_W - \mathrm{MI}_{SY}, \tag{3.133}$$

where MI_W is defined in Eq. (3.94) and MI_{SY} is defined in Eq. (3.130).

Note, that MI_W^p is not equivalent to the definition of $UI(W' : W \setminus A)$ given above, because $CI(W' : W; A)$ is different from MI_{SY} (compare Eq. (3.107) with Eq. (3.129)).

This concludes the discussion of morphological intelligence in the context of synergistic and unique information. In the next section, we discuss the quantifications of morphological intelligence, which are based on asking how much of the complexity of the world process could be in-sourced into the policy.

3.6 Concept Four: Quantifying Morphological Intelligence as In-Sourceable Computation

The fourth concept is the inversion of the widely used colloquial definition of morphological computation:

> Morphological Computation refers to the computation which is conducted by the body, that otherwise would have to be performed by the brain [10].

This definition can be understood in the following way: morphological computation refers to the computation that was somehow outsourced from the brain to the body. We invert this notion and declare that morphological intelligence is the computation that can be "*in-sourced*" from the morphology and its interactions with the environment back into the brain. This is in alignment with the definition of morphological intelligence as it was given in the first chapter of this book (see Definition 1.1), namely, the reduction of computational cost for the brain or controller. Given an observed behaviour, the concept discussed in this section asks the complementary question of how much more complex or costly the computation could be made without changing the observed behaviour.

As discussed previously (see Sect. 3.3.1), one way to estimate the complexity of a controller is the mutual information of the sensor state S and actuator state A:

$$I(A; S) = H(A) - H(A|S). \tag{3.134}$$

The one-step behaviour of a reactive system is described by the conditional probability distribution over the world states [85], i.e., $p(w'|w)$. Instead of asking for a maximum entropy model of $p(a|s)$, we are now asking for the minimum entropy model, i.e., the policy model with the most structure that produces the same behaviour $p(w'|w)$. This means we are looking for a policy π that has the highest mutual information under the constraint that the behaviour $p(w'|w)$ is preserved, denoted by

$$\pi^* = \operatorname{argmax}_\pi I(A; S)|_{p(w'|w)}. \tag{3.135}$$

Morphological intelligence, in this context, is then defined as the difference between the policy π^* with the highest possible complexity and the currently utilised policy π, given the observed behaviour:

$$\mathrm{MI_{IN}} := I^{\pi^*}(A; S) - I^\pi(A; S). \tag{3.136}$$

It can be interpreted as the amount of computation that can be "*in-sourced*" into the controller without affecting the behaviour.

Let us evaluate this method based on the example of the Passive Dynamic Walker [56–58], which is a mechanical structure that emulates the walking behaviour of humans if placed on a slope (see Fig. 1.2). We assume that the system is equipped with sensors that give an accurate representation of the body's state. Such a description could be the position and orientation of the hip in 3D as well as all the joint angles, foot contact information, etc. Next, we assume that the system can be equipped with actuators in each joint, such that the passive body dynamics is not altered. We can now compare the Passive Dynamic Walker as it was constructed with an equivalent system that is actively controlled.

For passive walking, the actuators are set into passive mode, which means that for all sensor states only one action is performed. Hence,

$$I(A; S) = H(A) - H(A|S) \leq H(A) \tag{3.137}$$
$$\leq \log |A| \tag{3.138}$$
$$= \log 1 \tag{3.139}$$
$$= 0. \tag{3.140}$$

Next, we estimate how much computation the information-theoretically most complex controller would require, with the constraint that the behaviour $p(w'|w)$ is preserved.

The most complex policy, that leads to the highest mutual information is the deterministic policy for the following reasons. For deterministic policies, the action A depends deterministically on the sensor S, and hence, the uncertainty about the action A is zero if the sensor state S is known, i.e., $H(A|S) = 0$. It then follows for deterministic policies that

$$I(A; S) = H(A) - H(A|S) \tag{3.141}$$
$$= H(A) \tag{3.142}$$
$$\leq \log |A|. \tag{3.143}$$

Inequality (3.143) follows from Eq. 2.61. It then follows that the amount of "*in-sourceable*" computation is

$$\text{MI}_{\text{IN}} = \log |\mathcal{A}| - I^{\pi}(A|S), \tag{3.144}$$

where $I^{\pi}(A|S)$ is the complexity of the policy that generated the observed behaviour.

Unfortunately, there is some arbitrariness to this solution. Reconsider the Passive Dynamic Walker as an example. The set of actions could be the desired angular positions for each leg and knee joint. In this case, $\log |\mathcal{A}|$ would be considerably large, because we have many values for $a \in \mathcal{A}$. Assuming that the maximal angle between the two legs is $60°$ and that we use a resolution of $1°$ for the control. This would mean that the maximally complex policy under these circumstances requires $\log 60 \approx 5.9$ bits. But the same system can also be controlled with forces, where $a \in \mathcal{A} = \{-f, 0, f\}$, where f is the force of the leg segments that results from the interaction with the environment, i.e., gravity. In this case, $\log 3 \approx 1.6$ bits. We could have also chosen a finer granularity for the angular control above (e.g. $0.5°$ steps), which would result in a higher value. This shows that there are some cases in which there is no canonical way to determine the most complex behaviour. The final concept in this chapter overcomes this arbitrariness by comparing the acting policy against a generally applicable controller, i.e., a generally applicable baseline behaviour.

3.7   Concept Five: Quantifying Morphological Intelligence as the Reduction of Computational Cost

The concept described in this section is inspired by Krakauer's notion of intelligence and stupidity [86, 87]. Krakauer defines intelligence as making a difficult problem easy, whereas stupidity is defined as making an easy problem difficult.

The concept is illustrated with a Rubik's cube. Any random behaviour will finally solve the cube, which means that each of the six sides will only show patches of the same colour. Any behaviour that solves the cube in fewer steps than the averaged random policy, will be called intelligent, whereas any behaviour which requires more steps compared to the averaged random behaviour is referred to as stupid (see Sect. 1.3 for a more detailed description).

Random behaviour is chosen as baseline behaviour because it does not require any strategy or memory. In other words, it is the behaviour that requires the least amount of knowledge about a problem and its solution that is able to solve almost any problem.

This concept of intelligence, stupidity, and the definition of a baseline behaviour is applied to morphological intelligence in the following way. We state that a morphology is intelligent if it makes the hard problem of control easier. This is equivalent to stating that the body reduces the computational cost (see Definition 1.1). Analogously, we would say that a morphology is stupid if it makes the problem of control more difficult or even impossible. The introduction (see Chap. 1) discusses several examples, of which running on uneven ground is briefly repeated here.

The elasticity of the leg's muscle-tendon system compensates for unevenness of the ground and reduces the amount of computation that the brain has to conduct to calculate feet placement. An example of a morphologically stupid system would a soft robot with unfavourable compliance. One example is a soft robot hand which is so soft that it can barely lift an object. Now the controller has to conduct additional computation to compensate for the undesired compliance, e.g., by predicting when the compliance will have a negative influence on the grasp and how this can be compensated by chaining the object is grasped (e.g. instead of grasping the object from the top one could roll the object into the hand to lift it). Morphological Intelligence and Morphological Stupidity for soft manipulation are discussed in Sect. 5.1.

The remaining question now is, what is the appropriate reference behaviour in the context of morphological intelligence. In Krakauer's definition, it was the random behaviour because it does not require any strategy or memory. We are interested in quantifying the effect of the body's physical properties and the body's interaction with the environment on the computational cost. Hence, the reference behaviour in this context should describe a system's behaviour without taking the physics into account.

In robotics, there is such a general and well-established method to describe the motion of any arbitrary system, without taking physics into account, which is known as inverse kinematics (see e.g. [88]). Inverse kinematic will be explained next based on the example of a hexapod (six-legged walking machine, see Fig. 3.14) with 2 DOF in each leg.

In general, forward and inverse kinematics are described by the following set of equations:

$$e = f(q) \qquad \text{[forward kinematics]} \qquad (3.145)$$

$$q = f^{-1}(e), \qquad \text{[inverse kinematics]} \qquad (3.146)$$

where $e \in \mathbb{R}^n$ is the pose (position and orientation) of the end effector or effectors and $q \in \mathbb{R}^m$ are the joint parameters. Forward kinematics means that we have a mathematical model f of the system and compute the end effector poses e as a function of the joint parameters q. Inverse kinematics means that we have a goal pose e for the end effectors and want to determine the joint parameters q that lead to this pose using the inverse of the robot model f^{-1}.

The Jacobian matrix

$$J(q) = \frac{\partial e(q)}{\partial q} \qquad (3.147)$$

describes the local movements of the end effectors as a function of local changes in the joint parameters. Hence, forward kinematics, i.e., the movement of the segments as a function of the joint parameters of any embodied system can be described in the following way:

$$\dot{e} = J(q)\,\dot{q}. \tag{3.148}$$

In the case of our hexapod example (see Fig. 3.14), Eq. (3.148) describes the motion of the feet as a function of the joint angles.

Controlling a system means that we want to know how to change the joint parameters such that the robot segments follow a specific trajectory. In the case of our hexapod (see Fig. 3.14) we want to know how to change the joint parameters q such that the system shows a specific walking behaviour (e.g. tripod walking pattern). This means that we need to calculate the inverse of the Jacobian matrix that describes the robots motion (see Eq. (3.147)). The inverse does not exist in general, because the Jacobian matrix shown in Eq. (3.147) is generally not a square matrix, i.e., the number of joint parameters m does not match the number of segment coordinates n. One possibility to calculate the inverse of a non-square matrix is the Moore-Penrose inverse, which leads to the following solution for the inverse kinematics problem:

$$\dot{q} = (J^T J)^{-1} J^T \dot{e}. \tag{3.149}$$

Equation (3.149) is the reference behaviour that we used to quantify Morphological Intelligence and Morphological Stupidity, analogous to Krakauer's random policy that was discussed in the context of the Rubik's cube above.

The random policy was chosen by Krakauer because it does not make use of any memory or strategy. Hence, any improvement with respect to the random policy is most likely includes either memory or strategy or both, i.e., includes some form of intelligent problem-solving. We are interested in the reduction of the computation cost as a result of the physical properties of the body and its interaction with the environment. Inverse kinematics does take the geometry of the body into account (Matrix f in Eq. (3.145) and in Eq. (3.146)) but ignores all physical contributions (e.g. inertia) to a behaviour. This means that in Eq. (3.149) we assumed that the motion of the end effectors $\dot{e}$ are only influenced by changes of the joint parameters $\dot{q}$. Inverse kinematics does also not take into account how body symmetries reduce the computational cost, as each part of the body is treated individually. Hence, any reduction of the computational cost with respect to inverse kinematics is the result of physical properties of the body (e.g. symmetry) and its interaction with the environment (e.g. inertia). We will make this explicit based on the example of a hexapod (see Fig. 3.14), which is discussed next.

Our example systems is a walking machine with six legs and a total of 12 degrees of freedom, i.e., each leg has two controlled joints. In the following paragraphs, we

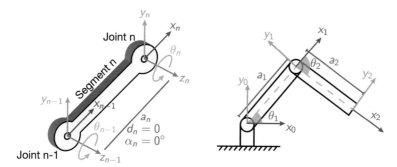

Fig. 3.13 *Visualisation of the Denavit-Hartenberg Notation.* The Denavit-Hartenberg notation follows three steps. Step 1: Naming of the segments and joints. Each segment receives a number between 1 and n, the basis is segment 0. Joint n connects segment $n - 1$ with segment n. Step 2: Defining the coordinate systems. Coordinate frame 1 is on joint 1 and parallel to the world coordinate frame. The coordinate frame of segment n is close to joint $n + 1$. Step 3: Orienting the coordinate frames. The z-axis of each coordinate frame is aligned with the joint axis (rotation or translation). The x-axis of coordinate frame n is directed to the origin of the coordinate frame $n + 1$. The y-axis results from the right-hand rule. For a detailed description, the reader is referred to standard literature on robotics (e.g. [88, 90]) or the original publication [89]. The parameters α, d, θ, and a describe the rotation and translation between two coordinate frames. They are explained in detail in the text below (see Eq. (3.150))

will first derive the mathematical representation of the robot (matrix f in Eq. (3.145)) using the Denavit-Hartenberg Notation [89], which requires homogeneous transformation matrices.

3.7.1 Denavit-Hartenberg Notation

In general, the location and orientation of a segment in three-dimensional space require six parameters. The first three parameters determine the position along the three cartesian axes, i.e., x, y, and z. The second three parameters determine the orientation around the three cartesian axes. They are often denoted by α, β, and γ (Euler-Notation). The Denavit-Hartenberg Notation (see Fig. 3.13 and [89]) is a method to reduce the number of parameters in a kinematic chain from six to four for each segment, by describing the pose of a segment in the local coordinate frame of the previous segment in the chain. The four parameters are translations along the x and z axes of the previous segment, denoted by d and a, and the rotations around the x and z axes of the previous segment, denoted by θ and α (see Fig. 3.13). The z-axis determines either the rotational or the translational axis of a joint (rotational or prismatic joint), which means that either d or θ are open for control.

The 4×4 homogeneous transformation matrix that describes the pose of segment j relative to segment i is then given by:

$$^{j}T_{i}(\theta, d, \alpha, a) = \text{Rot}(z_{i}, \theta_{i})\text{Trans}(z_{i}, d_{i})\text{Trans}(x_{i}, a_{i})\text{Rot}(x_{i}, \alpha_{i}) \qquad (3.150)$$

$$= \begin{pmatrix} \cos(\theta_{i}) & -\sin(\theta_{i}) & 0 & 0 \\ \sin(\theta_{i}) & \cos(\theta_{i}) & 0 & 0 \\ 0 & 0 & 1 & 0 \\ 0 & 0 & 0 & 1 \end{pmatrix} \begin{pmatrix} 1 & 0 & 0 & 0 \\ 0 & 1 & 0 & 0 \\ 0 & 0 & 1 & d_{i} \\ 0 & 0 & 0 & 1 \end{pmatrix} \begin{pmatrix} 1 & 0 & 0 & a_{i} \\ 0 & 1 & 0 & 0 \\ 0 & 0 & 1 & 0 \\ 0 & 0 & 0 & 1 \end{pmatrix}$$

$$\times \begin{pmatrix} 1 & 0 & 0 & 0 \\ 0 & \cos(\alpha_{i}) & -\sin(\alpha_{i}) & 0 \\ 0 & \sin(\alpha_{i}) & \cos(\alpha_{i}) & 0 \\ 0 & 0 & 0 & 1 \end{pmatrix}$$

$$= \begin{pmatrix} \cos(\theta_{i}) & -\sin(\theta_{i})\cos(\alpha_{i}) & \sin(\theta_{i})\sin(\alpha_{i}) & a_{i}\cos(\theta_{i}) \\ \sin(\theta_{i}) & \cos(\theta_{i})\cos(\alpha_{i}) & -\cos(\theta_{i})\sin(\alpha_{i}) & a_{i}\sin(\theta_{i}) \\ 0 & \sin(\alpha_{i}) & \cos(\alpha_{i}) & d_{i} \\ 0 & 0 & 0 & 1 \end{pmatrix}$$

How the coordinate systems are aligned in each joint with respect to the corresponding segments is briefly described in Fig. 3.13. For a detailed discussion, please see [88, 89].

In the following sections, we will use the homogeneous transformation matrix and the Denavit-Hartenberg Notation to derive the inverse kinematics to calculate the control parameters q_{t} for a tripod walking behaviour. The computational complexity of the derived solution will be compared to the complexity of the sine-wave central pattern generator (sine-wave CPG).

3.7.2   Denavit-Hartenberg for a Hexapod

In this section, we derive the equations for the inverse kinematics approach for a hexapod with 12 degrees of freedom (see Fig. 3.14). This section will close with a comparison of the inverse kinematics approach with a sine-wave CPG controller that was used e.g. in [85] to control such a robot. To simply the computation for the comparison, we only calculate the trajectories of the feet. The same method could be used to calculate the trajectories of each segment or joint. This would mean that the inverse kinematics must be applied iteratively along each kinematic chain, where kinematic chain refers to a chain of segments that are connected by motors (or some other form of actuation).

The task is to compute the 12 control parameters $q = (\theta_{i})_{i=1,2,...,12}$ for a tripod walking behaviour. The legs of the hexapod are morphologically equivalent. In particular, they have the same number of degrees of freedom and their local Denavit-Hartenberg parameters, i.e., angular and transitional offset of the segments, are the same. Hence, we can derive the equations for all leads simultaneously. The index k will refer to the k-th leg in each equation.

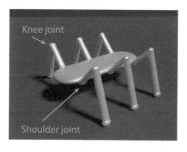
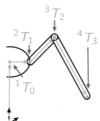
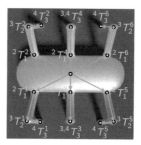

Fig. 3.14 *Visual representation of the transformation matrices for a hexapod.* ${}^{i}T_{j}^{k}$ *denotes the* transformation from coordinate frame j to i of the k-th leg, where T are 4×4 homogeneous transformation matrices (see text for details). Hence, ${}^{4}T_{0}^{1} = {}^{4}T_{3}^{1} \cdot {}^{3}T_{2}^{1} \cdot {}^{2}T_{1}^{1} \cdot {}^{1}T_{0}^{1}$ is the transformation from the base (centre of the main body) to the end-effector for the first leg

The position of the feet e as a function of the joint parameter q is given by the following equation:

$$e = T(q) = ({}^{4}T_{0}^{k}(q_{k}))_{k=1,2,...,6}, \tag{3.151}$$

where q_{k} are the joint parameters of the k-th leg, i.e. $q_{k} = (q_{k}^{1}, q_{k}^{2})$ (see Fig. 3.14), are the angular position of the first and second joint in each leg, $q = (q_{k})_{k=1,2,...,6}$, and ${}^{4}T_{0}^{k} = {}^{4}T_{3}^{1} \cdot {}^{3}T_{2}^{1} \cdot {}^{2}T_{1}^{1} \cdot {}^{1}T_{0}^{1}$ is the transformation from the base (centre of the main body) to the end-effector of the k-th leg.

To summarise this section, in order to produce a tripod walking behaviour for the hexapod shown in Fig. 3.14, we need to solve the following set of equations for a sequence of feet poses e^{t}, where $t = 1, 2, \ldots, T$ is the description of the tripod walking behaviour for one period:

$$T(q^{t}) = ({}^{4}T_{0}^{k}(q_{k}^{t}))_{k=1,2,...,6} \tag{3.152}$$

$$J(q^{t}) = \frac{\partial T(q^{t})}{\partial q^{t}} \tag{3.153}$$

$$\dot{q}^{t} = ((J(q^{t}))^{T} J(q^{t}))^{-1} (J(q^{t}))^{T} \dot{e}^{t}. \tag{3.154}$$

To ensure a tripod walking behaviour, Eq. (3.154) has to be solved in every time step. With respect to tripod walking, we define Eq. (3.154) as our baseline behaviour, analogous to the random behaviour that Krakauer defines as his baseline behaviour (see Sect. 1.3).

In Sect. 1.2.4, we discussed that there should be a quantifiable difference between the Passive Dynamic Walker and a ball rolling downhill. We believe that this concept results in a meaningful distinction between the Passive Dynamic Walker and the ball rolling downhill for the following reason. The ball's movement can be fully described by a translation and rotation, i.e., the current position of the ball on the slope and its current local rotation. The amount of computation required to solve the Denavit-Hartenberg equation for this movement is the amount of Morphological Intelligence, i.e., how much the computational load was reduced due to an exploitation of the environment (slope) and the body (ball). Equation (3.154) is significantly more complex for the Passive Dynamic, which has at least four joints that need to be parametrised. Hence, the reduction of the computation load is significantly higher for the Passive Dynamic Walker compared to the ball rolling downhill. This shows that this approach is able to distinguish between the Passive Dynamic walking and the ball rolling downhill in a meaningful way. In the next step, we will compare Eq. 3.154 for the hexapod with a control strategy that exploits the morphology.

We know that each leg has the same morphology and is connected to the body in the same way, i.e. the legs are parallel to each other and orthogonal to the main axis of the main body. This means that for a tripod walking behaviour, the control can be simplified in the following way. Legs 1, 5, and 3 can perform the same motion as well as legs, 2, 4, and 6 (see Fig. 3.14). Hence, we only have to calculate the joint angles for two legs (e.g. leg 1 and 2) and can copy the values to all the other legs (see hexapod example in YARS distribution [91, 92]). The full controller is then defined in the following way:

$$\theta_1 = \sin\left(t \cdot 2\frac{\pi}{50}\right) \qquad\qquad \theta_{7,9} = \theta_1 \qquad \theta_{3,5,11} = -\theta_1 \qquad (3.155)$$

$$\theta_2 = \sin\left(t \cdot 2\frac{\pi}{50} + 3\frac{\pi}{2}\right) \qquad \theta_{8,10} = \theta_2 \qquad \theta_{4,6,12} = -\theta_2 \qquad (3.156)$$

where t is the time.

For a more realistic tripod walking behaviour, in which there are clearly distinguishable stance and swing phases for each leg, the sine waves need to be modified, which means that the computations become more complex. Yet, the overall motion can still be produced by generating two phase-shifted control signals, which are then copied to the other legs. In both cases, the main point remains, which is that the exploitation of the morphology, i.e., the morphological equivalence of the legs and their attachment to the main body, allows to significantly reduce the amount of computation compared to the baseline behaviour (inverse kinematics, see Eq. (3.154)).

The question now is, how can this be quantified? The first two concepts, for example, will fail in this case, because the behaviour is fully deterministic, and hence the two conditional entropies $H(W'|W)$ and $H(W'|A)$, which determine the upper limit for MI_A and MI_W (see Eqs. (3.97) and (3.100)), will be close or equal to zero. This also holds true for the third concept, which measures the synergistic information that the current action A and the current world state W hold about the next world state W' that is not available if only one of the two random variables A or W is taken into account. In the case of the open-loop controller discussed above, almost all information about the behaviour can be extracted from knowledge about the current action A. Hence, the synergistic information will also be close to zero. The fourth concept compares the entropy of the actions chosen by the maximally complex and the currently applied policy, especially in the case of open-loop control, which does not access sensor states S. In the case described above, these two policies

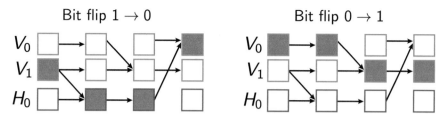

Fig. 3.15 *Visualisation of space cost and time step cost.* This plot shows the number of idempotent functions (time step cost) and hidden states (space) cost for a bit flip. Image is redrawn from [93]

are indistinguishable from the actions alone, hence, the measure would also estimate zero morphological intelligence.

There are currently three possible ways that are investigated to calculate morphological intelligence as the difference of the complexities of the current policy with respect to a reference policy. The first possibility is to transform both policies, the actually implemented (e.g. open-loop tripod walking behaviour, see Eq. (3.155) to Eq. (3.156)) and the reference policy (inverse kinematics) into assembly code. The computational cost can then be determined by counting the number of routines required for both approaches and subtracting these from each other. This approach depends heavily on the chosen set of assembly routines. A more fundamental approach is discussed by Wolpert et al. [93]. In their work, the authors describe two concepts, which they call *time step cost* and *state space cost* (or *space cost*). The concept is illustrated along with the example of a bit flip, which is shown in Fig. 3.15. In this case, we have a binary variable X, that can be in one of two visible states V_0 or V_1. To flip a bit, i.e., to find a function that implements $X = \text{NOT} Y$, we need one hidden state, here denoted by H_0. In this example, *space cost* refers to the number of required hidden states for an operation and *time stop cost* refers to the number of idempotent functions, which in this case are three (see Fig. 3.15). An idempotent function is a function that does not change the state of the system after it was applied for the first time. In Fig. 3.15, the first applied idempotent function maps $V_0 \rightarrow V_0$ and $V_1, V_2 \rightarrow V_2$. Applying it a second time will not change the values of the three states. A bit flip, therefore, has a *space cost* of one and a *time step cost* of three. This can be used to compare the computational costs of different policies in terms of *time step cost*, given that there are no limitations with respect to the *space cost*.

For the third method, we can interpret the visible states V_i as the observable behaviour of an embodied agent. In terms that we used previously to describe the sensorimotor loop (see Sect. 3.2), the visible states correspond to the world states $W_t, W_{t+1}, W_{t+2}, \ldots$, and the hidden states H_i correspond to the unity of the sensor, actuator, and controller states S_t, A_t, and C_t. The *space cost* corresponds to the complexity that the brain or controller may exploit to produce a certain behaviour. Complexity is here understood similarly to [85] in which it was used analogously to the number of hidden unit in a conditional restricted Boltzmann machine. One can now ask, how does the *time step cost* vary if there are limitations on the *space cost*? No limitations on the space cost mean that there are no limitations on the potential complexity of the idempotent functions, and hence, no limitations on the amount of computation. The lower bound for morphological intelligence is then related to the corresponding time step cost for the unlimited space cost setting. The question now is, how much can the space cost be limited while still preserving the behaviour of the system?

This concludes the presentation of the different concept to quantify morphological intelligence. The next chapter presents numerical analyses of those measures, that are currently computable. This concludes the concept presented in this section.

References

1. Pfeifer R, Iida F (2005) Morphological computation: connecting body, brain and environment. Japanese Sci Month 58(2):48–54
2. Paul C (2004) Morphology and computation. In: Proceedings of the international conference on the simulation of adaptive behaviour, Los Angeles, CA, USA, pp 33–38
3. Pfeifer R, Lungarella M, Iida F (2007) Self-organization, embodiment, and biologically inspired robotics. Science 318(5853):1088–1093
4. Hauser H, Ijspeert A, Füchslin RM, Pfeifer R, Maass W (2011) Towards a theoretical foundation for morphological computation with compliant bodies. Biol Cybern 105(5–6):355–370
5. Füchslin RM, Dzyakanchuk A, Flumini D, Hauser H, Hunt KJ, Luchsinger RH, Reller B, Scheidegger S, Walker R (2012) Morphological computation and morphological control: steps toward a formal theory and applications. Artif Life 19(1):9–34
6. Jaeger H, Haas H (2004) Harnessing nonlinearity: predicting chaotic systems and saving energy in wireless communication. Science 304(5667):78–80
7. Maass W, Natschläger T, Markram H (2002) Real-time computing without stable states: a new framework for neural computation based on perturbations. Neural Comput 14(11):2531–2560
8. Zhao Q, Nakajima K, Sumioka H, Hauser H, Pfeifer R (2013) Spine dynamics as a computational resource in spine-driven quadruped locomotion. In: 2013 IEEE/RSJ international conference on intelligent robots and systems, pp 1445–1451
9. Nakajima K, Hauser H, Kang R, Guglielmino E, Caldwell DG, Pfeifer R (2013b) Computing with a muscular-hydrostat system. In: 2013 IEEE international conference on robotics and automation, pp 1504–1511
10. Pfeifer R, Bongard JC (2006) How the body shapes the way we think: a new view of intelligence. The MIT Press (Bradford Books), Cambridge, MA
11. Hauser H, Corucci F (2017) Morphosis-taking morphological computation to the next level. Springer International Publishing, Cham, pp 117–122
12. Nurzaman SG, Yu X, Kim Y, Iida F (2014) Guided self-organization in a dynamic embodied system based on attractor selection mechanism. Entropy 16(5):2592–2610
13. Nurzaman SG, Yu X, Kim Y, Iida F (2015) Goal-directed multimodal locomotion through coupling between mechanical and attractor selection dynamics. Bioinspiration and Biomimetics 10(2):025004
14. Pfeifer R, Iida F, Gòmez G (2006) Morphological computation for adaptive behavior and cognition. Int Congr Ser 1291:22–29
15. Pfeifer R, Gómez G (2009) Morphological computation–connecting brain, body, and environment. Springer, Berlin, Heidelberg, pp 66–83
16. Clark A (1996) Being there: putting brain, body, and world together again. MIT Press, Cambridge, MA, USA
17. Zahedi K, Ay N (2013) Quantifying morphological computation. Entropy 15(5):1887–1915

18. Ay N, Zahedi K (2014) On the causal structure of the sensorimotor loop. In: Prokopenko M (ed) Guided self-organization: inception, emergence, complexity and computation, vol 9. Springer, pp 261–294
19. Ghazi-Zahedi K, Deimel R, Montúfar G, Wall V, Brock O (2017a) Morphological computation: the good, the bad, and the ugly. In: 2017 IEEE/RSJ International Conference on Intelligent Robots and Systems (IROS), pp 464–469
20. Ghazi-Zahedi K, Haeufle DF, Montufar GF, Schmitt S, Ay N (2016) Evaluating morphological computation in muscle and dc-motor driven models of hopping movements. Front Robot AI 3(42):
21. Jost J (2005) Dynamical systems. Springer
22. Strogatz SH (1994) Nonlinear dynamics and chaos. Addison-Wesley, Reading, MA, USA
23. Thomson JMT, Stewart B (2002) Nonlinear dynamics and chaos, 2nd edn. John Wiley and son LTD, New York, NY, USA
24. Meiss J (2007) Dynamical systems. Scholarpedia 2(2):1629
25. Iida F, Pfeifer R (2006) Sensing through body dynamics. Robot Auton Syst 54(8):631–640
26. Iida F, Gomez G, Pfeifer R (2005) Exploiting body dynamics for controlling a running quadruped robot. In: ICAR '05. Proceedings, 12th international conference on advanced robotics, pp 229–235
27. Iida F, Pfeifer R (2004) cheap rapid locomotion of a quadruped robot: self-stabilization of bounding gait. In: Proceedings of the international conference on intelligent autonomous systems, pp 642–649
28. Horsman C, Stepney S, Wagner RC, Kendon V (2014) When does a physical system compute? In Proceedings of the royal society a: mathematical, physical and engineering science 470(2169)
29. Jaeger H (2002a) Adaptive nonlinear system identification with echo state networks. In: Thrun S, Obermayer K (eds) Advances in neural information processing systems 15. MIT Press, Cambridge, MA, pp 593–600
30. Rückert EA, Neumann G (2013) Stochastic optimal control methods for investigating the power of morphological computation. Artif Life 19(1):115–131
31. Corucci F, Cheney N, Lipson H, Laschi C, Bongard J (2016) Material properties affect evolutions ability to exploit morphological computation in growing soft-bodied creatures. In: Proceedings of the artificial life conference 2016
32. Pfeifer R, Scheier C (1999) Understanding intelligence. MIT Press, Cambridge, MA, USA
33. Polani D (2011) An informational perspective on how the embodiment can relieve cognitive burden. In: Artificial life (ALIFE), 2011 IEEE symposium on, pp 78–85
34. Haeufle DFB, Günther M, Wunner G, Schmitt S (2014) Quantifying control effort of biological and technical movements: an information-entropy-based approach. Phys Rev E 89:012716
35. Klyubin A, Polani D, Nehaniv C (2004) Organization of the information flow in the perception-action loop of evolved agents. In: Proceedings of the 2004 NASA/DoD Conference on Evolvable Hardware, pp 177–180
36. Touchette H, Lloyd S (2004) Information-theoretic approach to the study of control systems. Phys A: Stat Mech Appl 331(1):140–172
37. Ay N, Zahedi K (2013) An information-theoretic approach to intention and deliberative decision-making of embodied systems. In: Advances in cognitive neurodynamics III, Springer, Heidelberg
38. von Förster H (1993) Wissen und Gewissen: Versuch einer Brücke, 1st edn. Suhrkamp-Taschenbuch Wissenschaft; 876, Suhrkamp, Frankfurt am Main, D
39. von Förster H (2003) Understanding understanding—essays on cybernetics and cognition. Springer, New York
40. Brooks RA (1986) A robust layered control system for a mobile robot. IEEE J Robot Autom 2(1):14–23
41. Brooks RA (1991a) Intelligence without reason. In: Myopoulos J, Reiter R (eds) Proceedings of the 12th international joint conference on artificial intelligence (IJCAI-91), Morgan Kaufmann publishers Inc.: San Mateo, CA, USA, Sydney, Australia, pp 569–595
42. Brooks RA (1991b) Intelligence without representation. Artif Intell 47(1–3):139–159

43. von Foerster H (2003) On self-organizing systems and their environments. Springer, New York, New York, NY, pp 1–19
44. Sung CH, Chuang JZ (2010) The cell biology of vision. J Cell Biol 190(6):953–963
45. Levick WR (1967) Receptive fields and trigger features of ganglion cells in the visual streak of the rabbit's retina. J Physiol 188(3):285–307
46. von Uexkuell J (1957) [1934]) A stroll through the worlds of animals and men. In: Schiller CH (ed) Instinctive behavior. International Universities Press, New York, pp 5–80
47. Zahedi K, Ay N, Der R (2010) Higher coordination with less control—a result of information maximization in the sensori-motor loop. Adapt Behav 18(3–4):338–355
48. Ay N, Löhr W (2015) The umwelt of an embodied agent–a measure-theoretic definition. Theory Biosci 134(3):105–116
49. Sutton RS, Barto AG (1998) Reinforcement learning: an Introduction. MIT Press
50. Ay N, Polani D (2008) Information flows in causal networks. Adv Complex Syst 11(1):17–41
51. Bauer H (1996) Probability Theory. De Gruyter studies in mathematics, Bod Third Party Titles
52. Shannon CE (1948) A mathematical theory of communication. Bell Syst Techn J 27:379–423
53. Pearl J (2000) Causality: models. Cambridge University Press, Reasoning and Inference
54. Aström K, Murray R (2010) Feedback systems: an introduction for scientists and engineers. Princeton University Press
55. Rivoire O, Leibler S (2011) The value of information for populations in varying environments. J Stat Phys 142(6):1124–1166
56. McGeer T (1990a) Passive dynamic walking. Int J Robot Res 9(2):62–82
57. McGeer T (1990b) Passive walking with knees. In: Robotics and automation, pp 1640–1645
58. Collins S, Ruina A, Tedrake R, Wisse M (2005) Efficient bipedal robots based on passive-dynamic walkers. Science 307(5712):1082–1085
59. Cover TM, Thomas JA (2006) Elements of information theory, vol 2nd. Wiley, Hoboken, New Jersey, USA
60. Schreiber T (2000) Measuring information transfer. Phys Rev Lett 85(2)
61. Bossomaier T, Barnett L, Harré M, Lizier JT (2016) An introduction to transfer entropy. Springer
62. Lizier JT (2014) The local information dynamics of distributed computation in complex systems. Springer
63. Bialek W, Tishby N (1999) Predictive information. https://arxiv.org/abs/cond-mat/9902341
64. Grassberger P (1986) Toward a quantitative theory of self-generated complexity. Int J Theor Phys 25(9):907–938
65. Polani D, Nehaniv C, Martinetz T, Kim JT (2006) Relevant information in optimized persistence vs. progeny strategies. In: Rocha LM, Bedau M, Floreano D, Goldstone R, Vespignani A, Yaeger L (eds) Proceedings artificial life X. MIT Press, Cambridge, MA, pp 337–343
66. Lungarella M, Pegors T, Bulwinkle D, Sporns O (2005b) Methods for quantifying the informational structure of sensory and motor data. Neuroinformatics 3:243–262
67. Williams PL, Beer RD (2010) Nonnegative decomposition of multivariate information. https://arxiv.org/abs/1004.2515
68. Bertschinger N, Rauh J, Olbrich E, Jost J, Ay N (2014) Quantifying unique information. Entropy 16(4):2161–2183
69. Griffith V, Koch C (2014) Quantifying synergistic mutual information. Springer, Berlin, Heidelberg, pp 159–190
70. Ghazi-Zahedi K, Rauh J (2015) Quantifying morphological computation based on an information decomposition of the sensorimotor loop. In: Proceedings of the 13th European conference on artificial life (ECAL 2015), pp 70–77
71. Ay N (2015) Information geometry on complexity and stochastic interaction. Entropy 17(4):2432–2458
72. Perrone P, Ay N (2016) Hierarchical quantification of synergy in channels. Front Robot AI 2:35
73. Harder M, Salge C, Polani D (2013) Bivariate measure of redundant information. Phys Rev E 87(1):012130

74. Griffith V, Chong EKP, James RG, Ellison CJ, Crutchfield JP (2014) Intersection information based on common randomness. Entropy 16(4):1985–2000
75. Bell AJ (2003) The co-information lattice. In: Proceedings of the fifth international workshop on independent component analysis and blind signal separation: ICA 2003
76. Makkeh A, Theis DO, Vicente R (2017) Bivariate partial information decomposition: the optimization perspective. Entropy 19(10)
77. Nowakowski PR (2017) Bodily processing: the role of morphological computation. Entropy 19(295)
78. SI Amari (2016) Information geometry and its applications. Springer
79. Oizumi M, Tsuchiya N, Si A (2016) Unified framework for information integration based on information geometry. Proc Natl Acad Sci 113(51):14817–14822
80. Kanwal MS, Grochow JA, Ay N (2017) Comparing information-theoretic measures of complexity in boltzmann machines. Entropy 19(7)
81. Darroch JN, Ratcliff D (1972) Generalized iterative scaling for log-linear models. Ann Math Stat 43(5):1470–1480
82. Csiszár I (1975) i-divergence geometry of probability distributions and minimization problems. Ann Probab 3(1):146–158
83. Ghazi-Zahedi K (2017a) Entropy++ GitHub Repository. http://github.com/kzahedi/entropy
84. Ghazi-Zahedi K (2017b) Go implementations of entropy measures. http://github.com/kzahedi/goent
85. Montúfar G, Ghazi-Zahedi K, Ay N (2015) A theory of cheap control in embodied systems. PLoS Comput Biol 11(9):e1004427
86. Kraukauer (2017) David Krakauer - q2. https://vimeo.com/125533384
87. Harris S (2016) Complexity & stupidity—a conversation with david krakauer. https://www.samharris.org/podcast/item/complexity-stupidity
88. Siciliano B, Khatib O (eds) (2008) Springer handbook of robotics, 2nd edn. Springer, Berlin, Heidelberg
89. Denavit J, Hartenberg RS (1955) A kinematic notation for lower-pair mechanisms based on matrices. Trans ASME E, J Appl Mech 22:215–221
90. McKerrow P (1991) Introduction to robotics. Addison-Wesley Pub Co., Sydney; Reading, Mass
91. Ghazi-Zahedi K (2016) YARS Github Repository. https://github.com/kzahedi/YARS
92. Zahedi K, von Twickel A, Pasemann F (2008) Yars: a physical 3d simulator for evolving controllers for real robots. In: Carpin S, Noda I, Pagello E, Reggiani M, von Stryk O (eds) SIMPAR 2008, Springer, LNAI 5325, pp 71—82
93. Wolpert DH, Kolchinsky A, Owen JA (2017) The minimal hidden computer needed to implement a visible computation. https://arxiv.org/abs/1708.08494

Chapter 4
Numerical Analysis of the Morphological Intelligence Quantifications

The proper method for inquiring after the properties of things is to deduce them from experiments.

Isaac Newton

The previous chapter introduced and discussed five different concepts to quantify morphological intelligence. The goal of this chapter is to investigate how the majority of these measures perform for different configurations of the sensorimotor loop. This means that we want to investigate how the measures perform, e.g. if the behaviour of the systems is fully determined by the action A or fully determined by the previous world state W and for all other possible configurations between these two.

Such an analysis is difficult to perform with data from real-world experiments for the following reason. In any real-world physical experiment, the causal dependency between the different variables in the sensorimotor loop cannot be manipulated at will. One example is the causal dependency of the current world state W on the next world state W'. This causal dependency reflects physics, i.e., the body and its interactions with the environment. In real-world experiments, this dependency is usually fixed or can only be varied within very narrow margins. Yet, to understand how the measures perform on real-world data, it is important to get a full picture of their performance, which means that it is desirable to vary the different mechanisms of the sensorimotor loop from complete independence to a fully deterministic coupling. A purely theoretical analysis may also not reveal the strength and weaknesses of the measures (as we will show below). Therefore, this chapter presents numerical analyses which are based on a parametrised model of the sensorimotor loop, more precisely, the sensorimotor loop for reactive systems (see Sect. 3.2 and Eq. (3.19) to Eq. (3.21)). The parametrised model, in which the strength of each causal coupling between random variables is determined by a single parameter, is presented in the next section. This will be followed by a presentation of the synergistic and unique measures ($UI(W' : W \setminus A)$, $CI(W' : W; A)$, $\mathrm{MI_{SY}}$, and $\mathrm{MI_W^p}$) because their results

© Springer Nature Switzerland AG 2019
K. Ghazi-Zahedi, *Morphological Intelligence*,
https://doi.org/10.1007/978-3-030-20621-5_4

Fig. 4.1 *Parametrised Model of the Sensorimotor Loop*. Please see the text below (see Sect. 4.1) and Eq. (4.2) to Eq. (4.8)

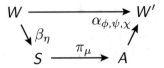

explain effects that are observable in the first two concepts. This chapter will close with a discussion of the presented results.

4.1 Parametrised Model of the Sensorimotor Loop

The causal model of the sensorimotor loop for reactive systems, as it is displayed in Fig. 4.1, shows that the joint distribution over all variables $p(w', w, s, a)$ is given by the product of three kernels and an input distribution, i.e.,

$$p(w', w, s, a) = p(w)\,\beta(s|w)\,\pi(a|s)\,\alpha(w'|w, a). \tag{4.1}$$

The parametrisation of the kernels and the input distribution are discussed in detail in the paragraphs below. It must be noted here that all random variables in this chapter use the same binary alphabet $w', w, a, s \in \{-1, 1\}$. The choice of -1 instead of 0 is a result of the particular parametrisation that is used here. The short explanation is that this particular choice of value ensures that the basis functions, that are defined by the random variables in the corresponding equations (see below), cover the entire space. For a full discussion of this argument, the reader is referred to [1].

4.1.1 Parametrisation of the Input Distribution $p(w)$

The presentation of the model starts with the input distribution $p(w)$, which is parametrised in the following way:

$$p_\tau(w) = \frac{e^{\tau w}}{\sum_{w^{\ddagger} \in \Omega} e^{\tau w^{\ddagger}}}. \tag{4.2}$$

In the remainder of this work, we will set $\tau = 0$, which results in a uniform distribution over the current world states, i.e., $p(w) = 1/2$ for $w \in \{-1, 1\}$. This ensures that the choice of $p(w)$ does not already impose information on the sensorimotor loop that could interfere with our analysis, e.g. by hiding other effects of interest.

4.1.2 Parametrisation of the Sensor Map $\beta(s|w)$

The sensor map $\beta(s|w)$ is given by:

$$\beta_\zeta(s|w) = \frac{e^{\zeta sw}}{\sum_{s^\ddagger} e^{\zeta s^\ddagger w}}. \tag{4.3}$$

As for the input distribution, this conditional distribution is controlled by a single parameter, namely $\zeta \in \mathbb{R}$. Three cases highlight how the distribution $\beta(s|w)$ changes with ζ. First, $\zeta = 0$ results in a uniform distribution, which is equivalent to the statement that S is independent of W. Second, for large values of ζ, i.e., $\zeta \to \infty$, Eq. (4.3) results in a deterministic map, which is best described by the following equation $\beta_{\zeta \to \infty}(s|w) = \delta_{sw}$, where δ_{sw} is the delta function, defined as

$$\delta_{xy} = \begin{cases} 1 \text{ if } x = y \\ 0 \text{ else} \end{cases}. \tag{4.4}$$

Third, for large negative values of $\zeta \to -\infty$, the sensor map $\beta(s|w)$ is again deterministic, but in a different way. It is best described by the following equation:

$$\beta_{\zeta \to -\infty}(s|w) = \begin{cases} 0 \text{ if } s = w \\ 1 \text{ else} \end{cases}. \tag{4.5}$$

In the binary case, the two configurations $\zeta \to -\infty$ and $\zeta \to \infty$ are equivalent which is why we will only plot for positive values of ζ in the numerical analysis below. For the same reason, we will also only plot for positive values of the parameters discussed for the following kernels.

4.1.3 Parametrisation of the Policy $\pi(a|s)$

The policy $\pi_\mu(a|s)$ is parametrised analogously to the sensor map (see Eq. (4.3)):

$$\pi_\mu(a|s) = \frac{e^{\mu as}}{\sum_{a^\ddagger} e^{\mu a^\ddagger s}}, \tag{4.6}$$

which means that the conditional distribution varies in dependence of $\mu \in \mathbb{R}$ in the same way it was described for the sensor map $\beta(s|w)$ above. The policy is random, i.e., the actions A are chosen independently of the current sensor state S, if $\mu = 0$, as Eq. (4.6) is then reduced to $p(a|s) = 1/2$. For large positive values of $\mu \to \infty$, the policy map is best described by the following equation:

$$\pi_{\mu \to \infty}(a|s) = \delta_{as}, \tag{4.7}$$

where δ_{as} is the Dirac measure (see Eq. (4.4)). For large negative values, the map is also deterministic. This case is omitted for the reason discussed in detail for the sensor map (see Sect. 4.1.2).

4.1.4 Parametrisation of the World Dynamics Kernel $\alpha(w'|w, a)$

The world dynamics kernel $\alpha(w'|w, a)$ is different, compared to the two previously presented conditional probability distributions, as it requires three parameters, namely ϕ, ψ, and χ. The first parameter $\phi \in \mathbb{R}$ determines the dependence of the next world state W' on the current world state W. In the terms that were used in the context of the information decomposition of the sensorimotor loop (see Sect. 3.4.1), ϕ controls the unique information that the current world state W contains about the next world state W' that is not contained in any way in A. Analogously, $\psi \in \mathbb{R}$ controls the unique information that the current action A contains about the next world state W' that is not available in W. Finally, $\chi \in \mathbb{R}$ corresponds to the synergistic information that the current world state W and action A contain about the next world state W'. As discussed earlier, values of zero refer to independence, i.e., no unique or synergistic information and large absolute values ($|\phi|$, $|\psi|$, $|\chi| \to \infty$) to a strong correlational or anti-correlational dependence of the corresponding variables. The parametrisation is given by the following equation:

$$\alpha_{\phi,\psi,\chi}(w'|w, a) = \frac{e^{\phi w'w + \psi w'a + \chi w'wa}}{\sum_{w^{\ddagger}} e^{\phi w^{\ddagger}w + \psi w^{\ddagger}a + \chi w^{\ddagger}wa}} \qquad (4.8)$$

To understand how the parameters act on the map, we will depict four exemplar cases, namely

1. Random case: $\phi = 0$, $\psi = 0$, $\chi = 0$
2. Passive Dynamic Walker: $\phi \to \infty$, $\psi = 0$, $\chi = 0$
3. Grid world with no embodiment: $\phi = 0$, $\psi =\to \infty$, $\chi = 0$
4. XOR: $\phi = 0$, $\psi = 0$, $\chi =\to \infty$

4.1.4.1 Case 1: The World is Purely Random

The first discussed case is given, when all three parameters are equal to zero, i.e., $\phi = 0$, $\psi = 0$, $\chi = 0$. Equation (4.8) is then given by:

$$\alpha_{\phi=0,\psi=0,\chi=0}(w'|w, a) = \frac{1}{2}. \qquad (4.9)$$

We will see in the following numerical analyses, that this case is actually not easy to determine in terms of morphological intelligence. It is clear that the next world state W' is not determined at all by the current action A, which means that morphological intelligence should be high. Yet, the next world state W' is also not determined by the previous world state W, which should lead to no measured morphological intelligence. This ambiguity will lead to contradicting results in this particular case for the different concepts and will be discussed below again.

4.1.4.2   Case 2: Passive Dynamic Walker

This second behaviour refers to the situation in which the next world state W' is only determined by the current world state W and fully independent of the current action A. This is the case in which the behaviour is only determined by the physics of the system, as it was described for the Passive Dynamic Walker (see Chap. 1). Formally, this case is defined by $\phi \to \infty$, $\psi = 0$, $\chi = 0$, which results in

$$\alpha_{\phi \to \infty, \psi=0, \chi=0}(w'|w, a) = \delta_{w',w}. \tag{4.10}$$

4.1.4.3   Case 3: Grid World with no Embodiment

In the third case that is selected for discussion, the next world state W' is only determined by the current action A, which means that the next world state is also fully independent of the current world state W. An example for such a system is a grid world, in which the agent can move without any restrictions. An illustrative example would be a chessboard in which the queen cannot only move into any direction but can jump to any position on the board. Formally, this is given by $\phi = 0$, $\psi =\to \infty$, $\chi = 0$, which is the binary case depicted above results in

$$\alpha_{\phi=0, \psi \to \infty, \chi=0}(w'|w, a) = \delta_{w',a}. \tag{4.11}$$

4.1.4.4   Case 4: XOR

The final case chosen for presentation is the situation in which the next world state W' depends on the current world state W *and* the current world A, which is also known as synergistic information of the current world and action state (see Sect. 3.5). This case is given by $\phi = 0$, $\psi = 0$, $\chi \to \infty$, which leads to the following equation for the world dynamics kernel:

$$\alpha_{\phi=0, \psi=0, \chi \to \infty}(w'|w, a) = w \text{XOR} a. \tag{4.12}$$

For the numerical analyses presented below, we will take all the cases discussed above into account. Hence, the numerical analyses will include results that span from random policies to deterministic policies (variations of μ), as well as variations that cover the four cases of the world dynamics kernel (variations of ψ, ϕ, and χ). Furthermore, the following sections will present numerical experiments for the different quantifications that were introduced in the previous section. Each experiment will be preceded with a brief explanation of the quantification and its defining equation (for a full explanation and discussion, the reader is referred to the previous chapter, see Chap. 3).

In all experiments presented below, we configure the sensor map $\beta(s|w)$ such that the sensor state S is a copy of the current world state W. This does not influence the generality of the results for the following reason. The sensor state S depends only on the world state W, which means that both variables can be merged into a single variable. In this case, the sensor map $\beta(s|w)$ would be incorporated into the policy $\pi(a|s)$, which would then become $\pi(a|w)$. The resulting model is mathematically equivalent to the *reactive* sensorimotor loop that includes the sensor state S. We chose a different path. Instead of removing S, we set the sensor map parameter ζ to a large value ($\zeta = 10$) which means that $\beta(s|w) \approx \delta_{sw}$. If not otherwise states, all experiments use the same values for the parameters:

$$\psi \in [0, 0.01, 0.02, \dots, 5] \qquad \chi \in [0, 1.25, 2.5, 3.75, 5] \qquad (4.13)$$

$$\phi \in [0, 0.01, 0.02, \dots, 5] \qquad \mu \in [0, 0.25, 0.5] \qquad (4.14)$$

$$\zeta = 10 \qquad (4.15)$$

This concludes the presentation of the parametrised model of the reactive sensorimotor loop and the parameters used within the remainder of this chapter.

Each analysis in the following will be discussed in four steps (see Fig. 4.2):

1. Plot A shown in the upper left corner of each figure, in which μ, $\chi = 0$, i.e., the action A is independent of the sensor state S, and there is no synergistic information between the current world state W, the current action A and the next world state W'.
2. Increasing synergistic parameter χ, while $\mu = 0$ (first row in each figure, plots B–E),
3. Increasing policy parameter μ, while $\chi = 0$ (first column in each figure, plots F and K),
4. Increasing policy and synergistic parameter, i.e., μ and χ are both increasing. This refers to the remaining plots, that were not yet discussed, from the upper left to the lower right corner (plots G–J and L–O).

Next, numerical results are presented for the measures which are based on synergistic and unique information.

MI$_{CA}$

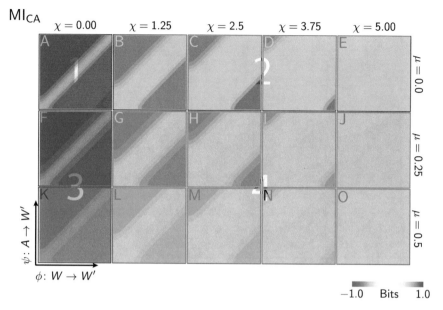

Fig. 4.2 *Order in which the plots are analysed.* This figure shows, based on the numerical results for MI$_{CA}$ (see Fig. 4.9), the order in which the plots are analysed. First, plot A is analysed (upper-left corner, $\mu, \chi = 0$. This plot highlights the effect of current world W and current action state A on $W′$. Second, plots B–E, are discussed, which show how synergistic information influences the measures. Third, plots F and K show how deterministic policies influence the measures if no synergistic information is present. Finally, forth, the remaining plots show the effects of all four parameters and can often be easily deduced from the previous three analytic steps.)

4.2 Numerical Results for CI(W′: W; A), UI(W′: W\A), MI$_{SY}$, and MI$_W^P$

Section 3.4 discussed the information decomposition of the conditional mutual information, in particular, its application to MI$_W$. Equation (3.114) shows that MI$_W$ is given by the sum of the unique information of the world states W, $W′$, i.e., $UI(W′ : W \setminus A)$, and the synergistic information $CI(W′ : W; A)$. Hence, MI$_W$ (and analogously MI$_A$) can be better understood if $UI(W′ : W \setminus A)$ and $CI(W′ : W; A)$ are investigated first. In this section, we will also include the analysis of MI$_{SY}$ and MI$_W^P$ as they are the synergistic and unique information based on [2]. Hence, this section will discuss the numerical analysis of the following four equations (see Chap. 3):

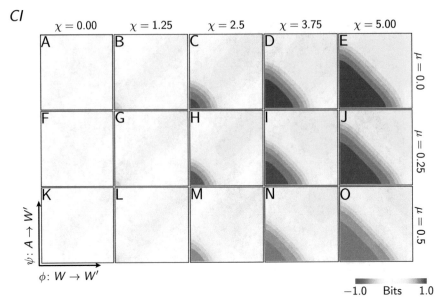

Fig. 4.3 *Numerical results for* $CI(W' : W; A)$. The colours in the contour plots are discretised for better visibility. In each plot, the lower-left corner represents the case in which unique information parameters are set to zero, i.e., $\psi = \phi = 0$. The parameter ψ is increased along with the y-axis and the parameter ϕ is increased along with the x-axis of each individual plot (see axes on the lower left). The upper right corner in each plot is given by $\phi = \phi = 5.0$. The synergistic parameter χ is increased from left to right over all plots. The χ value that is shown as column header at the top of this figure is valued for all three plots beneath it. Analogously, the policy parameter μ shown as row header on the right-hand side is valid for all five plots in the corresponding row. The colour refers to the amount of measured morphological intelligence, where white is 0, red is 1, and blue is -1 for all figures in this section. The plots shown in this figure are discussed in Sect. 4.2

$$CI(W' : W; A) = I(W' : W, A) - \min_{Q \in \Lambda_P} I_Q(W' : W, A) \qquad (4.16)$$

$$UI(W' : W \setminus A) = \min_{Q \in \Lambda_P} I_Q(W' : W | A) \qquad (4.17)$$

$$\mathrm{MI_{SY}} = D(p_{\mathrm{full}}(w'|w, a) || p_{\mathrm{split}}(w'|w, a)) \qquad (4.18)$$

$$\mathrm{MI_W^p} = I(W'; W | A) - \mathrm{MI_{SY}} \qquad (4.19)$$

Random policy, no synergistic information (μ, $\chi = 0$): The numerical results are shown in Figs. 4.3a, 4.4a, 4.5a, and 4.6a.

The synergistic measure $CI(W' : W; A)$ is low for all values of ψ and ϕ, which is expected for $\chi = 0$. Yet, there is a region along the diagonal $\psi \approx \phi$ in which this measure shows false positive values.

$UI(W' : W \setminus A)$ should increase with increasing parameter ϕ, as this is the only parameter that affects the causal dependency between W and W'. The desired behaviour for $UI(W' : W \setminus A)$ can only be seen for $\psi = 0$. If the coupling between

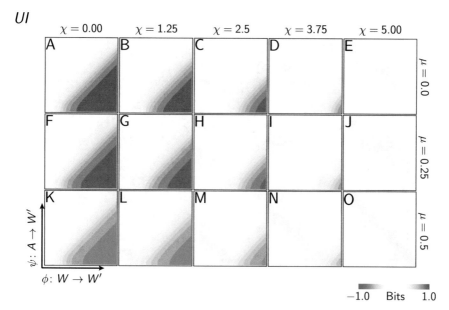

Fig. 4.4 *Numerical results for* $UI(W' : W \setminus A)$. The colours in the contour plots are discretised for better visibility. In each plot, the lower-left corner represents the case in which unique information parameters are set to zero, i.e., $\psi = \phi = 0$. The parameter ψ is increased along with the y-axis and the parameter ϕ is increased along with the x-axis of each individual plot (see axes on the lower left). The upper right corner in each plot is given by $\phi = \phi = 5.0$. The synergistic parameter χ is increased from left to right over all plots. The χ value that is shown as column header at the top of this figure is valued for all three plots beneath it. Analogously, the policy parameter μ shown as row header on the right-hand side is valid for all five plots in the corresponding row. The colour refers to the amount of measured morphological intelligence, where white is 0, red is 1, and blue is -1 for all figures in this section. The plots are discussed in Sect. 4.2

the action A and the next world state W' is increased, i.e., increasing parameter ψ (y-axis of plot A), it shadows the unique information $UI(W' : W \setminus A)$ in the sense that it requires a higher value for ϕ in order to detect unique information. This is seen in the plots because we only see positive values for $UI(W' : W \setminus A)$ if $\psi > \phi$, which should not be the case.

An important difference between $CI(W' : W; A)$ and MI$_{SY}$ is that MI$_{SY}$ does not show any false positives for $\chi, \mu = 0$ (compare Fig. 4.3a with Fig. 4.5a). This was already discussed in Sect. 3.5 and in [3] as an effect that results from excluding the feature that refers to the joint distribution in the measure from $CI(W' : W; A)$. $UI(W' : W \setminus A)$ and MI$_{W}^{P}$ show very similar behaviour, with the only difference, that the $UI(W' : W \setminus A)$ reaches maximal values for smaller values of ϕ.

Random policy, increasing synergy ($\mu = 0$, $\chi > 0$): The results are shown Figs. 4.3b–e, 4.4b–e, 4.5b–e, and 4.6b–e.

In this step, we investigate how an increasing synergistic parameter affects the results of the four measures. First, for $CI(W' : W; A)$, we see that positive synergistic

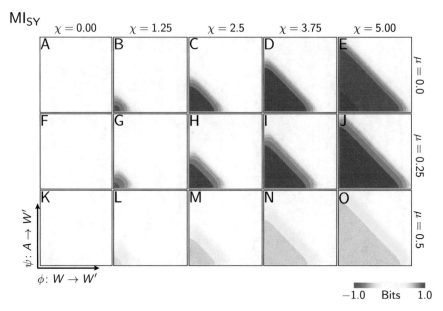

Fig. 4.5 *Numerical results for* MI_{SY}. The colours in the contour plots are discretised for better visibility. In each plot, the lower-left corner represents the case in which unique information parameters are set to zero, i.e., $\psi = \phi = 0$. The parameter ψ is increased along with the y-axis and the parameter ϕ is increased along with the x-axis of each individual plot (see axes on the lower left). The upper right corner in each plot is given by $\phi = \phi = 5.0$. The synergistic parameter χ is increased from left to right over all plots. The χ value that is shown as column header at the top of this figure is valued for all three plots beneath it. Analogously, the policy parameter μ shown as row header on the right-hand side is valid for all five plots in the corresponding row. The colour refers to the amount of measured morphological intelligence, where white is 0, red is 1, and blue is -1 for all figures in this section. The plots are discussed in Sect. 4.2

information for $\psi + \phi \lesssim \chi$ (lower-left corner in plots Fig. 4.3b–e). The region of false positives that were seen in the first step (discussed in the previous paragraph) increases in size and value for increasing χ (dark yellow regions in the centre of the plots).

The results for $UI(W' : W \setminus A)$ are as expected. The amount of unique information decreases with increasing synergistic parameter χ. Interestingly, and not directly visible through theoretical analysis, is the effect that at the border where $CI(W' : W; A)$ decreases from high values to 0, i.e., where $\psi + \phi \approx \chi$, we detect positive unique information. This is unexpected behaviour and must be classified as false positive values. At this point, the cause of this effect is not clear. To exclude the approximation methods as the source of this effect, different resolutions of the approximation method were evaluated.

For MI_{SY}, we see also high positive values for $\psi + \phi \lesssim \chi$, but surprisingly, values close to zero for $\chi > 0$ and $\psi + \phi > \chi$. This should not be the case, as synergistic information is present if $\chi > 0$. This must be classified as false negative values.

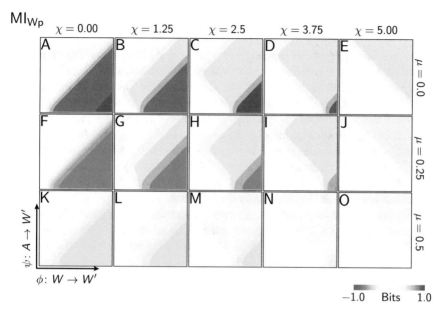

Fig. 4.6 *Numerical results for MI$_W^P$*. The colours in the contour plots are discretised for better visibility. In each plot, the lower-left corner represents the case in which unique information parameters are set to zero, i.e., $\psi = \phi = 0$. The parameter ψ is increased along with the y-axis and the parameter ϕ is increased along with the x-axis of each individual plot (see axes on the lower left). The upper right corner in each plot is given by $\phi = \phi = 5.0$. The synergistic parameter χ is increased from left to right over all plots. The χ value that is shown as column header at the top of this figure is valued for all three plots beneath it. Analogously, the policy parameter μ shown as row header on the right-hand side is valid for all five plots in the corresponding row. The colour refers to the amount of measured morphological intelligence, where white is 0, red is 1, and blue is −1 for all figures in this section. The plots are discussed in Sect. 4.2

Finally, the plots for MI$_W^P$ indicate that MI$_{SY}$ can be larger than the conditional mutual information $I(W′; W|A)$, because there are regions ($\phi + \psi \lesssim \chi$) for which MI$_W^P < 0$.

Non-random policy, no synergy ($\mu > 0, \chi = 0$): The results are shown Figs. 4.3f, k, 4.4f, k, 4.5f, k, and 4.6f, k.

The plots show that $CI(W′ : W; A)$ and MI$_{SY}$ are unaffected by an increasingly deterministic policy (if $\chi = 0$). This behaviour of the measures is expected because, in this setting, there is no synergistic information ($\chi = 0$) that can be affected by an increasingly deterministic policy $\mu > 0$. For the unique information measures $UI(W′ : W \setminus A)$ and MI$_W^P$, we see that the maximal values decrease for an increasing policy parameter μ. This can be explained intuitively in the following way. If the action A depends deterministically on the current world state W, then the information in W is no longer unique as it can also be gained through knowledge about A. The same argumentation also holds for the numerical results of MI$_W^P$ (shown in Fig. 4.6f, k).

Non-random policy, non-zero synergy ($\mu > 0$, $\chi > 0$): The results are shown Figs. 4.3g–j/l–o, 4.4g–j/l–o, 4.5g–j/l–o, and 4.6g–j/l–o.

For the two synergistic measures $CI(W' : W; A)$ and $\mathrm{MI}_{\mathrm{SY}}$, we see that the amount of detected synergistic information increases with an increase in χ, but it also decreases with an increase of μ. The explanation is the same as in the previous case. If the action A is more deterministically dependent on W, this means that (in the binary case) A is more likely a copy of W. If A is a copy of W, this means that there cannot be any synergistic information, because A and W are identical.

For the two unique measures, $UI(W' : W \setminus A)$ and $\mathrm{MI}_{\mathrm{W}}^{\mathrm{p}}$, we also see a combination of the two behaviours described above. With increasing χ, the amount of unique information decreases. This also happens for an increase of the policy parameter μ, i.e., an increasingly deterministic coupling of A and W.

Conclusion: Although theoretically very interesting, these measures do not quantify for application yet. The main reason is the false positive and false negative values of $CI(W' : W; A)$ and $UI(W' : W \setminus A)$. $\mathrm{MI}_{\mathrm{SY}}$ avoids the false positives but unfortunately shows false negatives, e.g. $\chi = 5.0$, $\phi = 5.0$, $\psi = 5.0$ (upper right plot). Here, we should see a positive value since $\chi > 0$, but $\mathrm{MI}_{\mathrm{SY}}$ does not detect any synergistic information. $\mathrm{MI}_{\mathrm{W}}^{\mathrm{p}}$ shows negative values for.

4.3 Numerical Results for $\mathrm{MI}_{\mathrm{A}}'$

MI_{A} was defined as the conditional mutual information $I(W'; A|W)$, i.e., as the information that the current action A contains about the next world state W' that is not already available in the current world state W (see Sect. 3.3 and Definition 3.4). It was argued that this measure must be inverted to be a quantification of morphological intelligence. This lead to the definition of $\mathrm{MI}_{\mathrm{A}}'$. In the binary case, it can be written as:

$$\mathrm{MI}_{\mathrm{A}}' = 1 - I(W'; A|W). \tag{4.20}$$

We chose to only visualise $\mathrm{MI}_{\mathrm{A}}'$ here, as it is the quantification that would be used to measure morphological intelligence. Furthermore, MI_{A} can be deduced directly from the plots, which means that adding the plots for MI_{A} will not result in additional insights. The numerical results for $\mathrm{MI}_{\mathrm{A}}'$ are shown in Fig. 4.7.

Random policy, no synergy ($\mu, \chi = 0$): The results are shown in Fig. 4.7a.

For $\phi > \psi$, i.e., if the coupling of the world state W and the next world state W' is stronger than the coupling of the current action A and the next world state W', $\mathrm{MI}_{\mathrm{A}}'$ results in positive values. Noteworthy is the case in which $\phi = 0$ and $\psi < \varepsilon$ (ψ small but positive). This is a significant difference to MI_{W} (see Fig. 4.11a), which is zero for these cases. This is one of the physically implausible situations described in the beginning of this chapter, because small values of ψ, ϕ, and χ mean that the next

MI′$_A$

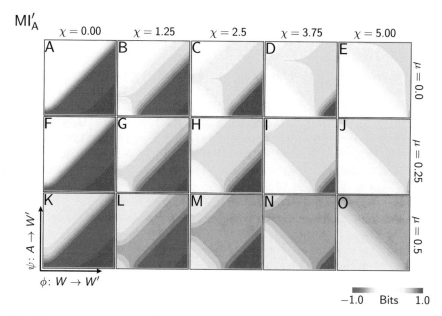

Fig. 4.7 *Numerical results for MI′$_A$.* The colours in the contour plots are discretised for better visibility. In each plot, the lower-left corner represents the case in which unique information parameters are set to zero, i.e., $\psi = \phi = 0$. The parameter ψ is increased along with the y-axis and the parameter ϕ is increased along with the x-axis of each individual plot (see axes on the lower left). The upper right corner in each plot is given by $\phi = \phi = 5.0$. The synergistic parameter χ is increased from left to right over all plots. The χ value that is shown as column header at the top of this figure is valued for all three plots beneath it. Analogously, the policy parameter μ shown as row header on the right-hand side is valid for all five plots in the corresponding row. The colour refers to the amount of measured morphological intelligence, where white is 0, red is 1, and blue is -1 for all figures in this section. The plots are discussed in Sect. 4.3

world state W' is mostly independent of the current world state W *and* the current action A. In some sense, on can argue that the world process is completely self-determined, which would justify a large value, and hence, the results are shown by MI′$_A$. On the other hand, one can also argue that the world process is also independent of its history, and hence, the results obtained by MI$_W$ are correct. This point will be discussed later in this chapter (see Sect. 4.9).

Random policy, non-zero synergy ($\mu = 0$, $\chi > 0$): The results are shown Fig. 4.7b–e.

This series of plots show that MI′$_A$ is zero for the regions at the lower left of each plot, which are given by $\phi + \psi \lesssim \chi$. At first, this seems counter-intuitive, as the conditional mutual information is the sum of the synergistic and unique information (see Eq. (3.106)), and hence, we would expect positive values when there is synergy. The reason is that MI′$_A$ is inverted with respect to MI$_A$, i.e.,

$$\mathrm{MI}'_A = 1 - I(W' : A|W) \qquad\qquad (4.21)$$
$$= 1 - UI(W' : A \setminus W) - CI(W' : A; W), \qquad (4.22)$$

which explains the regions with low values for increasing synergistic parameter χ.

Non-random policy, no synergy ($\mu > 0, \chi = 0$): The results are shown Fig. 4.7f, k.

We see that for increasing policy parameter μ, the amount of measured morphological intelligence increases overall, i.e., also for low values of ϕ. The reason is that this quantification measures morphological intelligence with respect to the effect of the action A has on the next world state W'. Increasing the policy parameter μ means that action A is more likely to be a copy of the world state W. Hence, the next world state W' can be fully predicted from knowledge about A alone. Consequently, MI_A decreases and MI'_A increases for an increasing policy parameter μ.

Non-random policy, non-zero synergy ($\mu > 0$, $\chi > 0$): The results are shown Fig. 4.7g–j/l–o.

These plots show the combination of the two previously described effects. Morphological intelligence decreases in the regions which are related to an increase of synergistic information, i.e., for which $\phi + \psi \lesssim \chi$, and the measured morphological intelligence increases with the policy parameter μ.

Conclusion: This measure will lead to false negative values if the investigated system has synergistic information (see Fig. 4.7).

4.4 Numerical Results for $\mathrm{MI_{MI}}$

$\mathrm{MI_{MI}}$ was defined as difference between the behaviour and policy complexity (see Definition 3.5),

$$\mathrm{MI_{MI}} = I(W'; W) - I(A; S). \qquad\qquad (4.23)$$

Random policy, no synergy ($\mu, \chi = 0$): The results are shown in Fig. 4.8a.

The case in which there is no synergistic information and the policy is random, $\mathrm{MI_{MI}}$ shows very similar behaviour to MI'_A, with the difference that it estimates zero morphological intelligence for $\phi \approx 0$.

Random policy, non-zero synergy ($\mu = 0$, $\chi > 0$): The results are shown in Fig. 4.8b–e. As for MI'_A, regions with positive synergistic information ($\psi + \phi \lesssim \chi$) are estimated with low morphological intelligence. The reason is seen in the following equations:

MI$_{\text{MI}}$

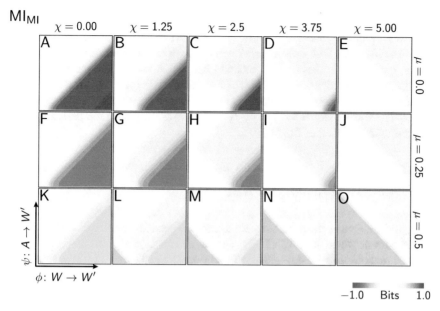

Fig. 4.8 *Numerical results for MI$_{MI}$.* The colours in the contour plots are discretised for better visibility. In each plot, the lower-left corner represents the case in which unique information parameters are set to zero, i.e., $\psi = \phi = 0$. The parameter ψ is increased along with the y-axis and the parameter ϕ is increased along with the x-axis of each individual plot (see axes on the lower left). The upper right corner in each plot is given by $\phi = \phi = 5.0$. The synergistic parameter χ is increased from left to right over all plots. The χ value that is shown as column header at the top of this figure is valued for all three plots beneath it. Analogously, the policy parameter μ shown as row header on the right-hand side is valid for all five plots in the corresponding row. The colour refers to the amount of measured morphological intelligence, where white is 0, red is 1, and blue is -1 for all figures in this section. The plots are discussed in Sect. 4.4

$$I(W'; A|W) = UI(W' : A \setminus W) + CI(W' : A; W) \tag{4.24}$$
$$= I(W'; W, A) - I(W'; W) \tag{4.25}$$
$$\Rightarrow I(W'; W) = I(W'; W, A) - UI(W' : A \setminus W) - CI(W' : A; W). \tag{4.26}$$
$$\text{MI}_{\text{MI}} = I(W'; W) - I(A; S) \tag{4.27}$$
$$= I(W'; W, A) - I(A; S) - UI(W' : A \setminus W) - CI(W' : A; W) \tag{4.28}$$

Non-random policy, no synergy ($\mu > 0$, $\chi = 0$): The results are shown in Fig. 4.8f, k. These plots show a clear difference to the plots of MI$'_{\text{A}}$. An increasing policy parameter μ reduces the amount of estimated morphological intelligence. This is expected, because the mutual information of the sensor and actuator state $I(A; S)$, i.e., the complexity of the policy, is subtracted from the complexity of the behaviour $I(W'; W)$. The plot shown in Fig. 4.8k shows a small region in which MI$_{\text{MI}}$ is negative ($\psi, \phi \approx 0$).

Non-random policy, no synergy ($\mu > 0$, $\chi > 0$): The results are shown in Fig. 4.8g–j/l–o. The plots are a combination of the two previously described effects, with the difference that we see larger regions with negative values of $\mathrm{MI}_{\mathrm{MI}}$. Intuitively, the reason is that both effects add up, which means that the reduction of morphological intelligence as a result of increasing synergistic information is combined with the reduction of morphological intelligence as a result of a more complex policy. The formal explanation, why $I(W'; W)$ decreases with increasing synergy was discussed above (see Eq. (4.26)).

4.5 Numerical Results for $\mathrm{MI}_{\mathrm{CA}}$

$\mathrm{MI}_{\mathrm{CA}}$ was defined as the difference of the *causal* dependence of next world state W' on the current world state W and the current action A (see Definition 3.6), which resulted in the following difference of conditional entropies

$$\mathrm{MI}_{\mathrm{CA}} = H(W'|A) - H(W'|W). \tag{4.29}$$

The numerical results are shown in Fig. 4.9.

Random policy, no synergy ($\mu, \chi = 0$): The results are shown in Fig. 4.9a. This plot shows that $\mathrm{MI}_{\mathrm{CA}}$ is significantly different with respect to the other measures discussed so far. This measure shows large negative values if μ, $\chi = 0$ and for $\psi < \phi$. In the terminology proposed in work, this measure is able to distinguish between morphological intelligence (positive values) and morphological stupidity (negative values).

Remaining plots: The results are shown in Fig. 4.9b–o. The remaining plots in Fig. 4.9 can be summarised in one paragraph, as they all show the same effect. With increasing synergy parameter χ and increasing policy parameter μ, the absolute value of the estimated morphological intelligence decreases, i.e., $\mathrm{MI}_{\mathrm{CA}} \to 0$.

4.6 Numerical Results for C_A

C_A is a quantification that is designed to work on variables, which are intrinsically available to the agent, i.e., it was designed to only operate on the current and next sensor states S, S' and the current action A (see Sect. 3.3.3). For the binary case, it is given by (see Definition 3.8):

$$C_A = 1 - \sum_{s,a} p(s, a) \sum_{s'} p(s'|\mathrm{do}(a)) \log \frac{p(s'|\mathrm{do}(a))}{p(s'|\mathrm{do}(s))} \tag{4.30}$$

MI$_{CA}$

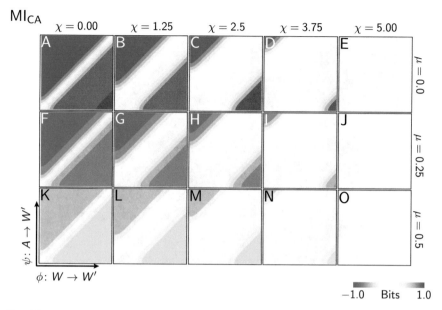

Fig. 4.9 *Numerical results for MI$_{CA}$.* The colours in the contour plots are discretised for better visibility. In each plot, the lower-left corner represents the case in which unique information parameters are set to zero, i.e., $\psi = \phi = 0$. The parameter ψ is increased along with the y-axis and the parameter ϕ is increased along with the x-axis of each individual plot (see axes on the lower left). The upper right corner in each plot is given by $\phi = \phi = 5.0$. The synergistic parameter χ is increased from left to right over all plots. The χ value that is shown as column header at the top of this figure is valued for all three plots beneath it. Analogously, the policy parameter μ shown as row header on the right-hand side is valid for all five plots in the corresponding row. The colour refers to the amount of measured morphological intelligence, where white is 0, red is 1, and blue is -1 for all figures in this section. The plots are discussed in Sect. 4.5

The numerical analysis of C_A requires the joint distribution:

$$p(w', w, s, s', a) = p(w)\beta(w|s)\pi(a|s)\alpha(w'|w, a)\beta(s'|w'), \qquad (4.31)$$

where the mechanisms are defined above (see Eqs. (4.2), (4.3), (4.6), and (4.8)).

Random policy, no synergy ($\mu, \chi = 0$): The results are shown in Fig. 4.10a. It seems, that this plot is identical to the plot shown for MI$'_A$ (see Fig. 4.7a), which indicates that the measure C_A, does capture the properties of the first concept. Hence, C_A is a good approximation for MI$'_A$ if the world states are not accessible.

Random policy, non-zero synergy ($\mu = 0$, $\chi > 0$): The results are shown in Fig. 4.10b–e. The plots indicate, that C_A can be understood as a superposition of the synergistic information measured by $CI(W' : W; A)$ and the unique information measured by $UI(W' : W \setminus A)$ (compare Fig. 4.10 with Figs. 4.3 and 4.4). This means that C_A is high if either $CI(W' : W; A)$ or $UI(W' : W \setminus A)$ is high.

C_A

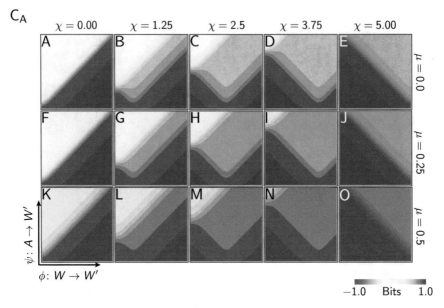

Fig. 4.10 *Numerical results for* C_A. The colours in the contour plots are discretised for better visibility. In each plot, the lower-left corner represents the case in which unique information parameters are set to zero, i.e., $\psi = \phi = 0$. The parameter ψ is increased along with the y-axis and the parameter ϕ is increased along with the x-axis of each individual plot (see axes on the lower left). The upper right corner in each plot is given by $\phi = \phi = 5.0$. The synergistic parameter χ is increased from left to right over all plots. The χ value that is shown as column header at the top of this figure is valued for all three plots beneath it. Analogously, the policy parameter μ shown as row header on the right-hand side is valid for all five plots in the corresponding row. The colour refers to the amount of measured morphological intelligence, where white is 0, red is 1, and blue is −1 for all figures in this section. The plots are discussed in Sect. 4.6

Non-random policy, no synergy ($\mu > 0$, $\chi = 0$): The results are shown in Fig. 4.10f, k. These plots show that C_A is very similar to MI'_A if $\chi = 0$. This was intended, as C_A was designed as a replacement for MI'_A, if only intrinsically accessible information is available. As for MI'_A, C_A estimates morphological intelligence overall higher if policy parameter μ is increased, even for low values of ψ and ϕ.

Non-random policy, non-zero synergy ($\mu > 0$, $\chi > 0$): The results are shown in Fig. 4.10g–j/l–o. The plots show a combination of the two previously described effects. Increasing the synergistic parameter leads to increased regions of positive values for morphological intelligence. This effect is enhanced by increasing the policy parameter μ.

4.7 Numerical Results for MI$_W$

MI$_W$ is the first quantification introduced in the second concept (see Sect. 3.4). It measures morphological intelligence as the effect of the current world state W on the next world state W, which leads to the following equation (see Definition 3.10):

$$\mathrm{MI_W} = I(W'; W | A). \tag{4.32}$$

The results are shown in Fig. 4.11.

Random policy, no synergy (μ, $\chi = 0$): The results are shown in Fig. 4.11a. As already briefly discussed for MI$_A$ (see Sect. 4.3), this plot highlights the difference between the concepts. MI$_A$ estimated high morphological intelligence for small values of ϕ and ψ, i.e., in cases in which the next world state W' is independent of the current action A and the current world state W. MI$_W$ estimates low morphological intelligence for these cases (lower left-hand side of Figs. 4.11a and 4.7a). This

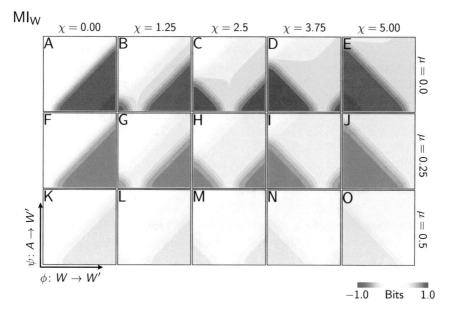

Fig. 4.11 *Numerical results for MI$_W$*. The colours in the contour plots are discretised for better visibility. In each plot, the lower-left corner represents the case in which unique information parameters are set to zero, i.e., $\psi = \phi = 0$. The parameter ψ is increased along with the y-axis and the parameter ϕ is increased along with the x-axis of each individual plot (see axes on the lower left). The upper right corner in each plot is given by $\phi = \phi = 5.0$. The synergistic parameter χ is increased from left to right over all plots. The χ value that is shown as column header at the top of this figure is valued for all three plots beneath it. Analogously, the policy parameter μ shown as row header on the right-hand side is valid for all five plots in the corresponding row. The colour refers to the amount of measured morphological intelligence, where white is 0, red is 1, and blue is −1 for all figures in this section. The plots are discussed in Sect. 4.7

difference will be discussed in the summary (see Sect. 4.9). Other than that, the behaviour of MI_W is equivalent to the behaviour of MI'_A for $\mu, \chi = 0$.

Random policy, non-zero synergy ($\mu = 0$, $\chi > 0$): The results are shown in Fig. 4.11b–e. The plots with $\mu = 0$ and increasing synergistic parameter χ show the second difference between the first an second concept. MI'_A (which is plotted in Fig. 4.7) is inverted, which lead to an estimation of zero morphological intelligence when synergistic information was present. This is different for MI_W. In this case, synergistic information is measured as morphological intelligence. This can also be seen formally (see Sect. 3.4):

$$MI_W = I(W'; W|A) \tag{4.33}$$
$$= UI(W' : W \setminus A) + CI(W' : W; A). \tag{4.34}$$

Non-random policy, no synergy ($\mu > 0$, $\chi = 0$): The results are shown in Fig. 4.11f, k. The third difference to the first concept is observable if the policy parameter μ is increased. As disused in Sect. 3.4, MI_W is zero for deterministic policies (see Eq. (3.104)). This can be observed in Figs. 4.11f, k. The maximal values of MI_W decrease for increasing policy parameter μ.

Non-random policy, no synergy ($\mu > 0$, $\chi > 0$): The results are shown in Fig. 4.11g–j/l–o. Increasing the synergistic parameter χ increases the range of values for ψ (coupling between action A and world state W') and ϕ (coupling between world states W and W') for which positive values of morphological intelligence are estimated. Increasing the policy parameter μ decreases the maximal measurable morphological intelligence, for the reasons mentioned above (upper limit of MI_W decreases with increasingly deterministic policy, see Sect. 3.4).

4.8 Numerical Results for C_W

C_W is an adaptation of MI_W that operates only on information that is intrinsically available to the agent, i.e., the current action A, the current sensor S, and the next sensor state S'. It is defined as (see Definition 3.14 in Sect. 3.4.2):

$$C_W = \sum_{s',s} p(s'|s)p(s) \log \frac{p(s'|s)}{\sum_a p(s'|a)p(a|s)}. \tag{4.35}$$

As for C_A, C_W also requires the joint distribution that includes the next sensor state, which generated in the following way

$$p(w', w, s, s', a) = p(w)\beta(w|s)\pi(a|s)\alpha(w'|w, a)\beta(s'|w'), \tag{4.36}$$

where the mechanisms are defined above (see Eqs. (4.2), (4.3), (4.6), and (4.8)).

The results of the numerical analysis are shown in Fig. 4.12.

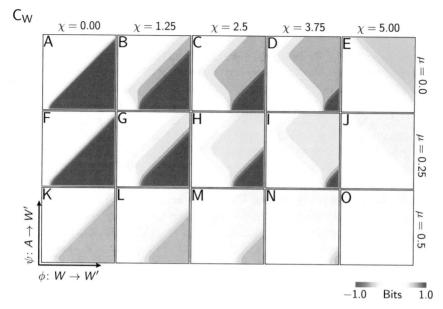

Fig. 4.12 *Numerical results for C_W.* The colours in the contour plots are discretised for better visibility. In each plot, the lower-left corner represents the case in which unique information parameters are set to zero, i.e., $\psi = \phi = 0$. The parameter ψ is increased along with the y-axis and the parameter ϕ is increased along with the x-axis of each individual plot (see axes on the lower left). The upper right corner in each plot is given by $\phi = \phi = 5.0$. The synergistic parameter χ is increased from left to right over all plots. The χ value that is shown as column header at the top of this figure is valued for all three plots beneath it. Analogously, the policy parameter μ shown as row header on the right-hand side is valid for all five plots in the corresponding row. The colour refers to the amount of measured morphological intelligence, where white is 0, red is 1, and blue is -1 for all figures in this section. The plots are discussed in Sect. 4.8

Random policy, no synergy $(\mu, \chi = 0)$: The results are shown in Fig. 4.12a. This plot shows a significant difference between MI_W and C_W. The transition from the low-value region (upper left-hand side of Fig. 4.11a and 4.12a) to regions with high values of morphological intelligence (lower right-hand side of Figs. 4.11a and 4.12a) are smoother for MI_W. By this, we mean that C_W is zero for $\psi > \phi$ (upper left-hand side of plot A) and one for $\psi < \phi$ (lower right-hand side of plot A). The transitional region between the two, i.e., the region in which $\phi \approx \psi$ is smaller compared to MI_W.

Remaining plots: The results are shown in Fig. 4.12b–o. The remaining plots can be summarised in the following way. As for MI_W, an increase in the policy parameter μ decreases the overall estimated morphological intelligence. The synergistic information as a negative effect on the estimated morphological intelligence (compare Fig. 4.3 with Fig. 4.12).

4.9 Summary

This chapter analysed the first three concepts of quantification morphological intelligence numerically. The fourth concept was omitted because it leads to trivial results. The results for the following quantifications were plotted: $UI(W' : W \setminus A)$, $CI(W' : W; A)$, MI_{SY}, MI_W^p, MI_A', MI_{MI}, MI_{CA}, C_A, MI_W, and C_W.

Based on the results obtained from the binary model of the sensorimotor loop, the measures for synergistic and unique information ($UI(W' : W \setminus A)$, $CI(W' : W; A)$, MI_{SY}, and MI_W^p) should not be used in applications, because they can lead to misleading results. By this we mean, that we either have false positives, false negatives, or even both.

There are three important aspects that should be considered if any of the remaining quantifications are chosen for application on real-world data. The first point was briefly discussed during the presentation of MI_A and MI_W (see also [4]). It is the question of how the quantifications should estimate morphological intelligence if the next world state W' is independent of the previous world state W. We believe that the range of values, for which the difference is observable, will not be relevant in practical application, because they relate to systems in which the next world state W' is almost completely independent of both, the action A and world state W. The second difference is the way synergistic information affects the measures. The quantifications MI_A', MI_{MI}, and C_W estimate low morphological intelligence, whenever $CI(W' : W; A)$ is large, i.e., there is detectable synergistic information. MI_{CA} is unaffected by synergy, and C_A, MI_W both detect synergy as morphological intelligence. From these results, the recommendation would be to use one of the following three MI_{CA}, C_A, or MI_W in applications. The last difference is the effect of an increasingly deterministic policy on the estimated morphological intelligence. The difference poses a fundamental question about the separation of body and brain, as the following example of the Braitenberg vehicle 3 ([5], see also Fig. 4.13) demonstrates. Braitenberg vehicle 3 is a *Gendankenexperiment* in which a simple wheel-driven robot (two actuated wheels) is equipped with light-sensitive sensors. The sensors are coupled with the wheel of the opposite side and may send the positive or negative signals to the wheels. A positive signal will accelerate the wheel, a negative value will decelerate the wheel. Hence, depending on the sign of the connection, the vehicles will either avoid or be attracted by a light source. The question here is, are these couplings part of the brain or part of the body? Both perspectives can be argued for. The connections between the sensors and motors can be understood as the connection in a (very simple) neural network. This would be in favour of MI_A' and low morphological intelligence. On the other hand, the connections are so simple, that they can also be understood as a mechanical coupling between sensors and actuators. This would be in favour of MI_W and high morphological intelligence.

This discussion shows that there are fundamental conceptual considerations that have to be taken into account if any of the measures are used for applications. This discussion also shows that we have so far not found the final answer to the question of quantifying morphological intelligence. We believe that the fifth concept, which

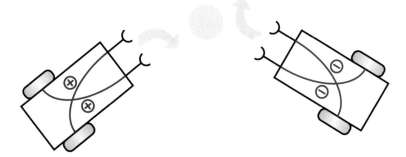

Fig. 4.13 *Braitenberg Vehicle 3a and 3b.* Images are redrawn from [5]. Both vehicles have cross-couplings between their two motors and two light-sensitive sensors. The only difference is that the vehicle on the left-hand side has positive coupling, i.e., a sensed light source will increase the wheel speed, whereas the vehicle on the right-hand side has negative couplings, i.e., a sensed light source will decrease the wheel speed. As a result, the vehicle on the right-hand side will accelerate and repeatedly collide with the light source. This behaviour could be interpreted as *aggression*. The vehicle on the right-hand side will explore the space until a light source is found and eventually stop in front of it. This can be interpreted as *affection* towards light. Both vehicles have very simple control structures and only differ in one particular property (the sign of the couplings) and yet one seeks and destroys light-sources, while the other shows affection towards them. This is an example which shows how very simple control structures can lead to behaviours that seem complex from an external point of view

measures morphological intelligence as the reduction of computational cost, can result in a quantification that does not suffer from these kinds of difficulties. This said, we do believe that the proposed measures can be useful in applications, as the next chapter demonstrates.

References

1. Ay N, Jost J, Le HV, Schwachhöfer L (2017) Information geometry. Springer
2. Ay N (2015) Information geometry on complexity and stochastic interaction. Entropy 17(4):2432–2458
3. Ghazi-Zahedi K, Langer C, Ay N (2017b) Morphological computation: synergy of body and brain. Entropy 19(9)
4. Zahedi K, Ay N (2013) Quantifying morphological computation. Entropy 15(5):1887–1915
5. Braitenberg V (1984) Vehicles. MIT Press, Cambridge MA

Chapter 5
Applications

> *It doesn't matter how beautiful your theory is, it doesn't matter*
> *how smart you are. If it doesn't agree with experiment, it's wrong*
> Richard P. Feynman

The first chapter discussed different forms of how the body reduces the computational cost for the brain, with examples from biology and robotics. Soft robotics and muscles were mentioned as examples. The second chapter introduced different ways to quantify morphological intelligence, and the previous chapter analysed these quantifications based on a parametrised model of the sensorimotor loop. In this work, we are not focussing only on a formal treatment, but also interested in the applicability of the measures on real data. From this type of analysis, it seems that MI_W, so far, is one of the best-suited candidates for applications. This was independently confirmed in previous publications [1–3]. This chapter presents our previous results and discusses them in the context of the recent developments that were presented in the previous chapters.

The first application is the quantification of morphological intelligence in the context of soft robot manipulation. In the second application, morphological intelligence is quantified on muscle models.

5.1 Quantifying Morphological Intelligence in Soft Manipulation

Soft robotics is a successful and relatively young branch of robotics. In many applications, softness leads to improved performance, robustness, and safety, while lowering manufacturing cost, increasing versatility, and simplifying control [4, 5]. In spite of these advantages, there currently is no systematic method for exploiting the benefits

© Springer Nature Switzerland AG 2019
K. Ghazi-Zahedi, *Morphological Intelligence*,
https://doi.org/10.1007/978-3-030-20621-5_5

Fig. 5.1 *Typical RBO Hand 2 grasp motion used to identify morphological intelligence and stupidity.* The grasp shown here is an instance of good morphological intelligence, which means that the compliance of the hand contributed to a firm grasp with a very simple controller. Morphological stupidity can be observed, if the object is not tightly held which leads an increased interaction of the hand and the object, e.g., by slowly slipping out of the hand. The goal of this section is to differentiate the two and make the results applicable in an automated design process

of softness in robot design. At the moment, human designers rely on experience and intuition to design competent soft robots.

The advantages of soft robots derive from the way their behaviour is generated. As with traditional robots, the behaviour of soft robots is affected by the control commands a robot receives. However, this control-based behaviour is modified through compliant interactions of the robot with its environment. These compliant interactions adapt the behaviour to a particular context, without the need for explicit control. It is therefore important to note that the behaviour of soft robots is *not* exclusively the result of control, it partially results from interactions of the robot's morphology with its environment. This latter part of the robot's behaviour, stemming from interactions, was formerly referred to as morphological computation [3]. In the context of this work, we now refer to it as morphological intelligence.

But not all body-environment interactions are good. The interactions between a soft robot and the environment can also be harmful, for example, if it un-does what control accomplished or simply causes failure. We previously called this *bad* or *ugly* morphological computation in [3]. In the context of this work, we would now call this *Morphological Stupidity*. To illustrate this with an example from soft manipulation: If hand-object interactions lead to the adaptation of a soft hand to the shape of an object that results in a good grasp (see Fig. 5.1), we call that morphological intelligence. If the compliance of the fingers lead to a less firm grasp, we consider this morphological stupidity as the controller has to compensate the harmful compliance by e.g. finding a different, more complex grasp. Both terms describe hand-object interactions, but only the former is desirable, while the latter is to be avoided.

An automated design of soft robots must minimize morphological stupidity and maximize morphological intelligence, relative to a particular task. In this section, we propose a method to identify morphological intelligence and stupidity from observed behaviour. If the observed behaviour can be represented in some high-dimensional space, our method identifies sub-spaces associated with morphological intelligence and sub-spaces associated with morphological stupidity. Such a criterion is a first and important step towards a quantitative design of soft robots.

We applied our method to soft grasping using an anthropomorphic robot hand based on pneumatic soft continuum actuators, known as RBO Hand 2 see also Fig. 5.2

[6]. Compared to a rigid manipulator, the design of a soft manipulator that is able to safely grasp a variety of objects is less obvious. To visualize this point, one can image a balloon slowly being filled with air. If otherwise unconstrained, the balloon will expand almost equally in all directions. For RBO Hand 2's fingers, this would be an undesired behaviour. Hence, a thread was carefully wrapped around the fingers in such a way that it allows and expansion of the dorsal (outer) side of the fingers, while it suppresses an expansion on the ventral (inner) side [6]. As a result, the hand closes when the air pressure is increased. Stated otherwise, certain degrees of freedom (DOF) of the fingers were restricted, while others where retained. The method proposed in this section is designed to automatically detect which DOFs should be suppressed and which should be enhanced in order to achieve a desired behaviour with minimal control.

The method consists of two steps. First, the components of the behaviour that can only be attributed to the physical hand-object interaction are extracted from the data. Second, the covariance of the DOFs is calculated. We show, that this form of dimensionality reduction can be used to distinguish between morphological intelligence and stupidity in a way that is applicable in an automated design process. The method is not limited to soft manipulation but can be applied to soft robotics in general.

5.1.1 Identifying Morphological Intelligence and Stupidity

The goal of the application presented first, is to find a method to systematically identify DOFs that contribute to a high grasp success with high morphological intelligence, i.e. that allow for a high versatility with a simplified control and to distinguish them from morphological stupidity (those DOF that impede good grasping). In other words, we want to enhance compliance that contributes to a good grasp and reduce compliance that reduces the grasp success. A scenario is an automated design process in simulation, in which a morphology is evaluated based on grasp attempts of several objects. In a purely evolutionary setting, the morphological parameters, e.g. the stiffness of the DOFs, would be open to random modifications. Soft manipulators inherently have many DOFs, which renders such an approach barely practical. Instead, a method to extract the characteristics of a grasp behaviour is required that can be used to guide an automated optimization process.

In the following sections, we first motivate covariance matrices as a method of dimensionality reduction and a way to extract the characteristics of motions. The calculations are performed on pre-processed data that only contain the hand's motion which results from the interaction of the hand with the object (see Sect. 5.1.3). To avoid artefacts, only a portion of the recorded grasps was used, which is described the last segment of this section on soft robotics.

5.1.2 Characterising Behaviours by Their Covariance

This section describes covariance matrices as a method of dimensionality reduction that captures the characteristics of grasps, and hence, can be used to distinguish between morphological intelligence and stupidity. This is motivated by other work e.g. [7], which showed that covariance matrices can be used to obtain a meaningful hierarchical clustering of, e.g. humanoid behaviours. In this particular example, different crawling behaviours were clustered closer together based on the covariances. The distance between clusters reflected the difference in the behaviours, e.g. crawling versus climbing out of a pit.

The hypothesis is that large covariance coefficients which are shared among experiments with high grasp success relate to useful compliance (morphological intelligence) and should be reinforced during the design process. Likewise, large coefficients shared among unsuccessful grasps should be suppressed, because they indicate harmful compliance (morphological stupidity). We will discuss this in more detail further below (see Sect. 5.1.10). We first describe how we calculate the covariance matrix of a grasp, before we discuss how the data are pre-processed. The covariance matrix of a data set $B \in \mathbb{R}^{T \times N}$ is given by

$$C(B) = (c_{ij})_{i,j=1,2,...,N} = \frac{1}{T} \sum_{t=1}^{T} (B_{t,i} - \bar{B}_i)(B_{t,j} - \bar{B}_j),$$

where $B_{t,j}$ refers to the entry in the t-th row and j-th column of the matrix B, and $\bar{B}_j = \frac{1}{T} \sum_{t=1}^{T} B_{t,j}$ is the mean value of the j-th column. In the context of this work, the matrix B contains the x, y, z values of each coordinate frame of the simulated hand (see Fig. 5.3) over time. The index j refers to each coordinate, hence $B_{t,1}, t = 1, \ldots, T$ is the recorded data for x_1 for an entire grasp. This means that the covariance coefficient c_{36} contains the covariance of the first and second frame's z coordinate. Hence, a large positive coefficient c_{36} would mean that the movements along the z-axis of the first and second frame should be highly correlated for a successful grasp. In an automated setting, the coupling of these two frames' movements would be enforced by increasing the stiffness along these DOFs.

5.1.3 Extracting Motions That Only Result from Body-Environment Interactions

We are interested in the interaction of a soft manipulator with an object. Therefore, we pre-process the data before calculating the covariance matrices in the following way. For each combination of a RBO Hand and controller (see below, Sect. 5.1.7), we first record the prescriptive behaviour, which is the hand's movement without any graspable object present in the scene. We refer to this data set as B_p. The record-

Fig. 5.2 *RBO Hand 2*. Left-hand side: Actual robot performing two different grasps that both result from the same control signal. The images show how the hand adapts to the object. Right-hand side: Simulation model of the hand

ing of the grasp itself is denoted by B_g. The element-wise difference of these two behaviours describes movements of the coordinate frames that are the result of the hand's interaction with the object and it is denoted by $B = B_g - B_p$. Hence, the covariance matrix $C(B)$ contains information about correlations of movements that only relate to the soft manipulator's compliance.

5.1.4 Avoiding Artefacts

A potential concern is that all recordings in which the object was dropped early or not grasped at all, will not differ significantly from the prescriptive behaviour, and hence $B = B_g - B_p$ will be mostly zero. This can be avoided if only a fraction of the time steps are taken into account, which is why we evaluated different time frames in our analysis. We chose to present the results based on 75 time steps (excluding the 10 approaching time steps, see Sect. 5.1.7) to avoid a clustering based on grasp success versus grasp failure as well as a clustering only reflecting the object's shape.

The next section describes the clustering method that was used to visualize the similarities of different grasps based on their covariance.

5.1.5 Visualising Clusters of Grasp Behaviours

We used t-SNE [8] to obtain a two dimensional representation of the covariance matrices C for visualization. This method, t-Distributed Stochastic Neighbor Embedding, constructs pairwise similarities of the input data and visualizes them in n dimensions, where $n = 2$ is chosen in this work. These visualizations capture the local structure of the data, while also revealing global structure such as the presence of clusters at several scales. The results section will show the obtained clusters coloured with the

Fig. 5.3 *Display of the different hand morphologies and objects used during the experiments.* Left-hand side: different hand morphologies and the set of objects used during the experiments. Right-hand side: Illustration of the model used to simulate soft hands: The trihedra indicate discrete links, adjacent ones are connected by passively compliant ball joints (purple trihedra) in between. The surface mesh is attached to the frames, and is used to compute collision and model contact

grasp success (explained next) and MI_W (see Sect. 3.4), but this information was not used during the clustering itself.

5.1.6 Determining Grasp Success

The grasp success is determined by the average distance of the object to the hand during the last 10 time steps of each recorded behaviour. The distance is measured between the object's geometric centre and the coordinate frame located at the first frame of the second finger (see Fig. 5.3). The object size is then subtracted from the measured distance to ensure comparability of the results for all objects. It was discussed above, that the covariance matrices were only calculated on a subset of the recorded data. This is not the case for the grasp distance, which was always estimated on the last 10 time steps of each full recording.

The next section discusses how the data were acquired.

5.1.7 RBO Hand Grasp Simulations

We used a simulation of the RBO Hand in order to obtain a large data set of fully observed grasping behaviours (see Fig. 5.1). Recent improvements to simulation algorithms [9] enable us to simulate complete grasp attempts in near real-time. The simulator is implemented with the SOFA framework [10] and relies on its Compliant module [9]. In this setup, soft hands are modelled as a tree of Cosserat beams (i.e. kinematic chains with ball joints), to which a collision surface is attached [11]. Figure 5.2 (right-hand side) shows the actual simulation model. Large trihedra indicate the links (beam elements), purple trihedra the joint location between two adjacent links. The surrounding wire-frame is the collision mesh, which is attached to the links via linear blend skinning and follow the motion of the actuator's backbone. Mechanical parameters are computed using a recently published model [6].

Using simulation makes it especially easy to obtain motion data, e.g. fingertip frame motion. In addition, simulations can be run in parallel, which results in a much larger data set to conduct the investigation on. Another advantage of using simulation is that the hand morphology can be changed easily. We created two variations of the RBO Hand 2 (see Fig. 5.3, left-hand side). Experiments were conducted with all three hands. These three hands were combined with three distinct motion primitives that implement variations of the surface-constrained grasp (see Fig. 5.1). This yields nine hand-controller combinations to compare. In the morphology domain, the spread between the four fingers was varied, and in the control domain the roll angle of the wrist during the grasping motion was varied in $\{-15°, 0°, 15°\}$.

5.1.8 Simulation Data

In total, the simulated hand consists of 32 coordinate frames (see Fig. 5.3). The dynamics of each finger (and thumb) are modelled by five coordinate frames. The remaining 7 coordinate frames are distributed along the palm and used to control the wrist and arm motion. For each frame, the simulator records the pose, of which only the 3D positional data are used in this work. The underlying assumption is that the orientation of the coordinate frames can be reconstructed from the x, y, z coordinates of consecutive frames. Furthermore, as we are interested in analysing grasping, we transformed all coordinate frames into the wrist frame, which means that the state of the hand is given by 31 coordinate frames, and hence, by a vector $b_t \in \mathbb{R}^{93}$.

We recorded the data for different hand designs, controllers, objects, and object's initial position (see Fig. 5.3 and previous section). The data include nine different hand-controller configurations, eight different objects, and finally, 27 different initial positions (x, y, θ) for each object, resulting in total 1944 recorded behaviours.

Each grasp is recorded for 300 time steps with a step width of 0.01 s and can be divided into four phases: 1. approaching of the object (hand moves downwards for 10 time steps), 2. grasping (30 time steps), 3. lifting (30 time steps), and finally, 4. evaluation of the grasp stability (remaining time steps). Hence, each grasp is captured in the matrix $B_g = \mathbb{R}^{300 \times 93}$.

5.1.9 Results

We present the covariance matrices of the grasps clustered with t-SNE (see Sect. 5.1.5) and coloured by grasp success, morphological intelligence measured by MI_W (see Sect. 3.4 and Definition 3.10), object type, and object initial position. This allows us to understand what kind of information is stored in the covariance matrices and if it can be used to distinguish between morphological intelligence and stupidity. For the

Fig. 5.4 *Clustering of grasp behaviours, coloured by grasp distance*. Smaller values (blue) correspond to better grasping The instances within the grey ellipse are those of interest, with high grasp success and high morphological intelligence

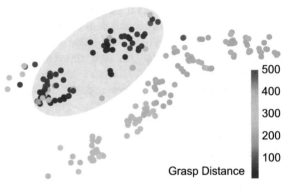

Fig. 5.5 *Clustering of grasp behaviours, coloured by MI$_W$*. Larger values (red) relate to better grasping. The instances within the Gray ellipse are those of interest, with high grasp success and high morphological intelligence

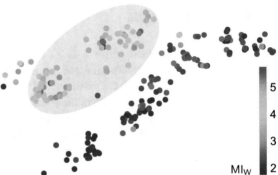

sake of clarity of the discussion, we decided to only plot and discuss the results for one RBO Hand and one controller as the results hold for every other combination too.

The main results are shown in Figs. 5.4 and 5.5. Figure 5.4 shows the clusters coloured by grasp success and Fig. 5.5 shows the clusters coloured by MI$_W$. By comparing the two plots, it is seen that the two large clusters resulting from the covariance matrices, can be very well explained by good grasping with high morphological intelligence (highlighted with grey ellipse in background) and unsuccessful grasps.

To verify that the clustering cannot be equally well explained by the object shape or object initial position, we also plot the clusters coloured by these two parameters. Figures 5.6 and 5.7 show that neither the object's shape nor its initial position can be used to fully explain the clustering. It also seems that the two sub-clusters of the cluster with high grasp success and high morphological intelligence can partly be explained by the object's shape. This means that, as can be expected, the covariance matrices not only contain information about correlations of DOFs that contribute to good grasping with high morphological intelligence, but also some information about the grasped object. The object's initial position (compare Fig. 5.6 with Fig. 5.7) is least informative in describing the two major clusters.

Fig. 5.6 *Clustering of grasp behaviours, coloured by object type*. This figure shows that there is some form of distinction possible based on the shape of the grasped object. Yet, the colouring by object type does not explain successful versus unsuccessful grasping and high versus morphological stupidity (compare with Figs. 5.4 and 5.5)

Fig. 5.7 *Clustering of grasp behaviours, coloured by object initial pose*. This figure shows that the clustering cannot be explained by the object's initial position

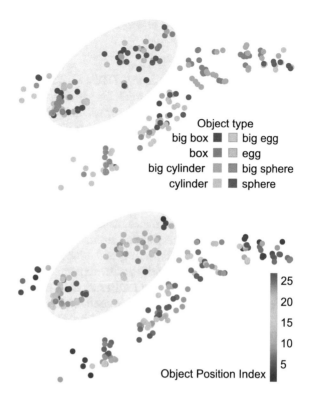

5.1.10 Discussion

The results (see Figs. 5.4, 5.5, 5.6, and 5.7) showed that a clustering based on the covariance matrices of the difference between grasping and its corresponding prescriptive behaviour leads to clusters which are best explained by high grasp success and high morphological intelligence (based on the quantification MI_W). The local relation of the instances reflect the similarities of the corresponding covariance matrices, which is what was referred to as sub-spaces of morphological intelligence and stupidity in the introduction.

If we look at representative examples of covariance matrices from sub-spaces with high density, two conclusions can be drawn. First, the matrices have a regular structure, which means that there is a high regularity in how the DOFs interact. Second, successful grasps have stronger positive covariance coefficients and weaker negative coefficients compared to less successful grasps. To summarize, we can identify coefficients that relate to morphological intelligence and stupidity in clusters with high density. Together with the regularity of the matrices, this suggests that the covariance coefficients can be used in an automated design process to guide modifications of the morphology (which is discussed next).

5.1.11   Covariance Coefficients Can Guide an Automatic Design Process

Positive coefficients correspond to DOFs which increase and decrease together. Negative coefficients correspond to DOFs which act reversely, i.e. if one increases, the other decreases. The latter could correspond to a finger movement, in which one of the coordinate frames moves upwards (positive z-movement) while the other coordinate frame moves downwards (negative z-movement). This would be an example of compliance that un-does what the controller tried to achieve. As the hand closes (e.g. negative z-movement of one the finger's coordinate frame) the "harmful" compliance of the finger (e.g. positive z-movement of another coordinate frame) prohibits a firm grasp.

A comparison the covariance matrices reveals that stronger negative coefficients correspond to less successful and stronger positive correlations correspond to successful grasping. Given the density of the sub-spaces (compare Fig. 5.4 with Fig. 5.5), it should be possible to identify positive coefficients which are most dominant over all successful grasps. The related DOFs should be enforced, e.g. by increasing the stiffness between them. Strong positive coefficients that are dominant in all unsuccessful grasps mean that the relative motion of the related DOFs should be softened. Negative coefficient can be used analogously. Hence, the dimensionality reduction based on covariance matrices can support a systematic modification of the morphological properties (e.g. stiffness) of a soft robot in an automated design process.

5.1.12   Conclusions

Currently, the success of a soft robot design relies on the expertise and intuition of its designer. The reason is that, up to this point, there is no systematic way to analyse how the compliance of a soft robot contributed to a desired behaviour. Such a systematic approach is the first required step in an automated design process of soft robots. This work is the first to propose such a method and to discuss how it could potentially be used to guide an automated design process.

We described physical processes which result from the interaction of the soft materials with the environment and are beneficial as morphological intelligence. Naturally, in soft robots, there are body-environment interactions that are harmful, which means that they might render the actions sent to the robot worthless. We referred to this form of compliance as morphological stupidity. Hence, for an automated design process of soft robots, we need a systematic way to identify morphological intelligence and stupidity based on observations of the robot's interaction with its environment.

For this purpose, we conducted a series of grasp experiments, with variations of RBO Hand 2 shape, object, object initial position, and controller. We showed that the covariance matrices calculated on the difference of grasping to prescriptive behaviour contain information that allows us to distinguish between morphological intelligence and stupidity based on observations alone. We discussed how the covariance coefficients relate to the compliance of the corresponding degrees of freedom and how the coefficients can be used to guide an automated design process.

The next step is to evaluate the proposed method in an automated process to design a soft manipulator for a specific task. To this point, we are able to identify morphological intelligence and stupidity, i.e. to distinguish DOFs which support the desired behaviour form those which harm the success. The final step is to use this information in a simulated set-up to modify the parameters of, e.g. a simulated hand. In particular, the co-variance coefficients of grasps in clusters with high density (small differences between the C-matrices), that also have high morphological intelligence and a good grasp, will be used to modify the softness/stiffness of RBO Hand 2's DOFs. This is the topic of currently ongoing research.

5.2 Quantifying Morphological Intelligence in Muscle Models

For biological systems, energy efficiency and adaptivity are important and evolutionary advantages. A strong indication for the importance of energy efficiency is given by the fact that the human brain accounts for only 2% of the body mass but is responsible for 20% of the entire energy consumption [12]. The energy consumption is also remarkably constant [13]. Under the assumption that the acquisition of energy was not a trivial task throughout most of the evolution of humans, on can safely conclude that as much computation as possible has been outsourced to the embodiment. When hunting prey or escaping from a predator, adaptivity to irregularities in the environment through morphological properties is a clear evolutionary advantage. Therefore, morphological intelligence may be a driving force in evolution.

In biological systems, movements are typically generated by muscles. Several simulation studies have shown that the muscles' typical non-linear contraction dynamics can be exploited in movement generation with very simple control strategies [14]. Muscles improve movement stability in comparison to torque driven models [15] or simplified linearised muscle models for an overview see [16]. Muscles also reduce the influence of the controller on the actual kinematics (they can act as a low-pass filter). This means that the hopping kinematics of the system is more pre-determined with non-linear muscle characteristics than with simplified linear muscle characteristics [16]. And finally, in hopping movements, muscles reduce the control effort (amount of information required to control the movement) by a factor of approximately 20 in comparison to a DC-motor driven movement [17].

The main contribution of this section is to evaluate two measures of morphological intelligence on biologically realistic hopping models. Based on our previous findings (see above), we hypothesize that morphological intelligence is higher in hopping movements driven by a non-linear muscle compared to those driven by a simplified linear muscle or a DC-motor. Furthermore, our experiments show that

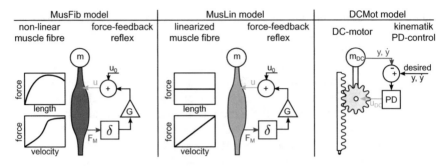

Fig. 5.8 *Hopping models.* The image is taken from [2]. The MusFib model considers the non-linear contraction dynamics of active muscle fibres and is driven by a mono-synaptic force-feedback reflex. The MusLin model only differs in the contraction dynamics, where the force-length relation is neglected and the force-velocity relation is approximated linearly. The DCMot model generates the leg force with a DC-motor. It is controlled by a proportional-differential controller (PD), enforcing the desired trajectory. The desired trajectory is the recorded trajectory from the MusFib Model. The sensor signals are shown as blue arrows, the actuator control signals are shown as green arrows. In case of the muscle models, the sensor signal is the muscle force F_M and the actuator control signal is the neural muscle stimulation u. In case of the DC-motor model, the sensor signals are the position and velocity of the mass, and the actuator control signal is the motor armature voltage u_{DC}

a point-wise analysis of morphological intelligence for the different models leads to insights, which cannot be gained from the averaged measures alone. Finally, we provide detailed instructions on how to apply these measures to robotic systems and to computer simulations, including MATLAB®, C++ code, and the data used in this Sect. [18].

5.2.1 Definition of Muscle and DC-Motor Models

In this section, we describe the three hopping models that were used in the experiments (see Fig. 5.8). In a reduced model, hopping motions can be described by a one-dimensional differential equation [19]:

$$m\ddot{y} = -mg + \begin{cases} 0 & y > l_0 \quad \text{flight phase} \\ F_L & y \le l_0 \quad \text{ground contact} \end{cases}, \tag{5.1}$$

where the point mass $m = 80\,\text{kg}$ represents the total mass of the hopper which is accelerated by the gravitational force ($g = -9.81\,\text{m/s}^2$) in negative y-direction. An opposing leg force F_L in positive y-direction can act only during ground contact ($y \le l_0 = 1\,\text{m}$). Hopping motions are then characterized by alternating flight and stance phases. In this section, we investigated three different models for the leg force. All models have in common, that the leg force depends on a control signal $u(t)$ and the system state $y(t)$, $\dot{y}(t)$: $F_L = F_L(u(t), y(t), \dot{y}(t))$, meaning, that the force modulation partially depends on the controller output $u(t)$ and partially on the dynamic characteristics, or material properties of the actuator. The control parameters of all three models were adjusted to generate the same periodic hopping height of $\max(y(t)) = 1.070\,\text{m}$. All models were implemented in MATLAB® Simulink™ (Ver2014b) and solved with ode45 Dormand-Prince variable time step solver with absolute and relative tolerances of 10^{-12}. To evaluate and compare the results of the models, a time-discrete output with constant sampling frequency is required. Therefore, the quasi-continuous variable time step signals generated by the ode45 solver are not adequate. To generate the desired output at 1 kHz sampling frequency, we used the Simulink built-in feature to generate desired output only. This way, the solver decreased step-size below 1 ms if required for precision, however, Simulink would only output the data for 1 kHz sampling frequency. This is similar to measuring a continuous physical system with a discrete time sensor. The models were solved for $T = 8\,\text{s}$.

5.2.1.1 Muscle-Fibre Model (MusFib)

A biological muscle generates its active force in muscle fibres whose contraction dynamics are well studied. It was found that the contraction dynamics are qualitatively and quantitatively (with some normalizations) very similar across muscles of all sizes and across many species. In the MusFib model, the leg force is modelled to incorporate the active muscle fibres' contraction dynamics. The model has been motivated and described in detail elsewhere [16, 17, 19]. In a nutshell, the material properties of the muscle fibres are characterized by two terms modulating the leg force

$$F_{L,\text{MusFib}} = q(t) F_{\text{fib}}(l_M, \dot{l}_M). \tag{5.2}$$

The first term $q(t)$ represents the muscle activity. The activity depends on the neural stimulation $u(t)$ of the muscle $0.001 \le u(t) \le 1$ and is governed by biochemical processes modelled as a first-order ODE called activation dynamics

$$\dot{q}(t) = \frac{1}{\tau}\left(u(t) - q(t)\right), \tag{5.3}$$

with the time constant $\tau = 10\,\text{ms}$. The second term in Eq. (5.2) F_{fib} considers the force-length and force-velocity relation of biological muscle fibres. It is a function of the system state, i.e., the muscle length $l_M = y$ and muscle contraction velocity $\dot{l}_M = \dot{y}$ during ground contact $y \le l_0$ and constant $l_M = l_0$ $\dot{l}_M = 0$ during flight $y > l_0$:

$$F_{\text{fib}} = F_{\max} \cdot \exp\left(-c \left| \frac{l_M - l_{\text{opt}}}{l_{\text{opt}} w} \right|^3 \right) \tag{5.4}$$

$$\times \begin{cases} \frac{\dot{l}_{M,\max} + \dot{l}_M}{\dot{l}_{M,\max} - K \dot{l}_M} & \dot{l}_M > 0 \\ N + (N-1)\frac{\dot{l}_{M,\max} - \dot{l}_M}{-7.56 K \dot{l}_M - \dot{l}_{M,\max}} & \dot{l}_M \leq 0 \end{cases}.$$

Here we use a maximum isometric muscle force $F_{\max} = 2.5\,\text{kN}$, an optimal muscle length $l_{\text{opt}} = 0.9\,\text{m}$, force-length parameters $w = 0.45\,\text{m}$ and $c = 30$, and force-velocity parameters $\dot{l}_{\max} = -3.5\,\text{ms}^{-1}$, $K = 1.5$, and $N = 1.5$ [19].

In this model, periodic hopping is generated with a controller representing a mono-synaptic force-feedback. The neural muscle stimulation

$$u(t) = G \cdot F_{L,\text{MusFib}}(t - \delta) + u_0 \tag{5.5}$$

is based on the time delayed ($\delta = 15\,\text{ms}$) muscle fibre force $F_{L,\text{MusFib}}$. Please note that this delay corresponds to the biophysical time delay due to the signal propagation velocity of neurons. The feedback gain is $G = 2.4/F_{\max}$ and the stimulation at touch down $u_0 = 0.027$.

This model neither considers leg geometry nor tendon elasticity and is therefore the simplest hopping model with muscle-fibre-like contraction dynamics. The model output was the world state $w(t) = (y(t), \dot{y}(t), \ddot{y}(t))$, the sensor state $s(t) = F_{L,\text{MusFib}}(t)$, and the neural control command $a(t) = u(t)$. For this model, these are the values that the random variables W, S, and A take at each time step.

5.2.1.2   Linearised Muscle-Fibre Model (MusLin)

This model differs from the model MusFib only in the representation of the force-length-velocity relation, i.e., $F_{L,\text{MusLin}} = q(t)F_{\text{lin}}(\dot{l}_M)$ (see Eq. (5.4)). More precisely, the force-length relation is neglected and the force-velocity relation is approximated linearly

$$F_{\text{lin}} = 1 \cdot (1 - \mu \dot{l}_M), \tag{5.6}$$

with $\mu = 0.25\,\text{m/s}$. Feedback gain $G = 0.8/F_{\max}$ and stimulation at touch down $u_0 = 0.19$ were chosen to achieve the same hopping height as the MusFib model.

5.2.1.3   DC-Motor Model (DCMot)

An approach to mimic biological movement in a technical system (robot) is to track recorded kinematic trajectories with electric motors and a PD-control approach. The DCMot model implements this approach slightly modified from [17]. The leg force generated by the DC-motor was modelled as

$$F_{L,\text{DCMot}} = \gamma T_{DC} = \gamma k_T I_{DC}, \tag{5.7}$$

where $k_T = 0.126\,\text{Nm/A}$ is the motor constant, I_{DC} the current through the motor windings, $\gamma = 100 : 1$ the ratio of an ideal gear translating the rotational torque T_{DC} and movement $\dot{\varphi}(t) = \gamma \dot{y}(t)$ of the motor to the translational leg force and movement required for hopping. The electrical characteristics of the motor can be modelled as

$$\dot{I}_{DC} = \frac{1}{L}\left(u_{DC} - k_T \gamma \dot{y}(t) - R I_{DC}\right), \tag{5.8}$$

where $-48\,\text{V} \le u_{DC} \le 48\,\text{V}$ is the armature voltage (control signal), $R = 7.19\Omega$ the resistance, and $L = 1.6\,\text{mH}$ the inductance of the motor windings. The motor parameters were taken from a commercially available DC-motor commonly used in robotics applications (Maxon EC-max 40, nominal Torque $T_{\text{nominal}} = 0.212\,\text{Nm}$). As this relatively small motor would not be able to lift the same mass, the body mass was adapted to guarantee comparable accelerations

$$m_{DC} = \frac{\gamma T_{\text{nominal}}}{F_{\text{max}}} m = 0.68\,\text{kg}. \tag{5.9}$$

The controller implemented in this technical model is a standard PD-controller. The controller tries to minimize the error between a desired kinematic trajectory ($y_{\text{des}}(t)$ and $\dot{y}_{\text{des}}(t)$) and the actual position and velocity ($y(t)$ and $\dot{y}(t)$) by adapting the motor voltage:

$$u_{DC}(t) = K_P(y_{\text{des}}(t) - y(t)) + K_D(\dot{y}_{\text{des}}(t) - \dot{y}(t)). \tag{5.10}$$

Here, the feedback gains are $K_P = 5000\,\text{V/m}$ and $K_D = 500\,\text{Vs/m}$. The desired trajectory during ground contact was taken from the periodic hopping trajectory of the MusFib model ($y_{\text{des}}(t) = y_{\text{MusFib}}(t)$ and $\dot{y}_{\text{des}}(t) = \dot{y}_{\text{MusFib}}(t)$).

This model is the simplest implementation of negative feedback control that allows to enforce a desired hopping trajectory on a technical system. The model output was the world state $w(t) = (y(t), \dot{y}(t), \ddot{y}(t))$, the sensor state $s(t) = (y(t), \dot{y}(t))$, and the actuator control command $a(t) = u_{DC}(t)$.

The DC-motor model is based on a Simulink model provided by Roger Aarenstrup.[1]

5.2.2 Experiments

This section discusses the experiments that were conducted with the hopping models and the pre-processing of the data. At this stage, the measures operate on discrete state spaces (see Chap. 3). Hence, the data was discretised in the following way. To

---

[1] http://in.mathworks.com/matlabcentral/fileexchange/11829-dc-motor-model.

Table 5.1 Numerical results for MI_W and MI_{MI}. The results for MI_W show that the non-linear muscle model MusFib contributes on average about 7.219 bits to the hopping behaviour, which is significantly more that the linear muscle model MusLin or the DC motor model (4.975 bits and 4.960 bits). The results are consistent for MI_{MI} although the linear muscle MusLin performs slightly better with respect to morphological intelligence compared to the DC motor model

	Muscle fibre model MusFib (bits)	Linearised muscle model MusLin (bits)	DC motor model DCMot (bits)
MI_W	7.219	4.975	4.960
MI_{MI}	7.310	5.153	4.990

ensure the comparability of the results, the domain (range of values) for each variable (e.g. the position y) was calculated over all hopping models. Then, the data of each variable was discretised into 300 values (bins). Different binning resolutions were evaluated and the most stable results were found for more than 100 bins. Finding the optimal binning resolution is a problem of itself and beyond the scope of this work. In practice, however, a reasonable binning can be found by increasing the binning until further increase has little influence on the outcome of the measures.

The possible range of actuator values are different for the motor and muscle models. For the muscle models, the values are in the unit interval, i.e., $a(t) \in [0, 1]$, whereas the values for the motor can have higher values (see above). Hence, to ensure comparability, we normalized the actions of the motor to the unit interval before they were discretised.

The hopping models are deterministic, which means that only a few hopping cycles are necessary to estimate the required probability distributions. To ensure comparability of the results, we parametrized the hopping models to achieve the same hopping height.

5.2.3 Results

The following paragraphs discuss the findings based on the averaged results presented in Table 5.1 first and then follow with a discussion of the point-wise results presented in Fig. 5.9.

All three models generated a similar movement, i.e., periodic hopping with a hopping height of 1.07 m. However, the control signals and the trajectories of the centre of mass vary between models (Fig. 5.11). Therefore, the computed values for the morphological intelligence vary between the models for both quantification methods (Table 5.1). Compared to the complex muscle fibre model (MusFib), the linearised muscle model (MusLin) and the technical DC-motor model (DCMot) result in significantly lower values of morphological intelligence ($\approx$ 30% less, see Table 5.1).

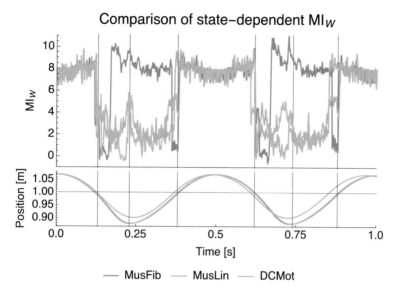

Fig. 5.9 *Point-wise evaluation of* MI_W *on the hopping models.* The upper plot shows the point-wise MI_W on the three hopping models. The lower plot visualizes the hopping position. The red line separates stance and flight phases. The plots only show a small fraction of the recorded data. The full data is shown in Fig. 5.11. For better readability, all the plots for morphological intelligence are smoothed with a moving average of block size 5. The plots show that morphological intelligence is high during the flight phase for all systems, i.e., whenever the behaviour of the system is only governed by gravity. Only for the MusFib, morphological intelligence is highest when the muscle is contracting. The other two models show low morphological intelligence when the muscle is contracting and in general when the hopper touches the ground

To analyse the differences between the models in more detail, we plotted the point-wise morphological intelligence. Fig. 5.9 shows the values of MI_W for each state of the models during two hopping cycles. We chose to discuss MI_{MI} only, because the corresponding values of MI_{MI} are very similar to those of MI_W, and hence, a discussion of the point-wise MI_{MI} will not provide any additional insights. This is why the numerical simulations in the previous chapter are so important. They showed a clear difference between MI_W and MI_{MI}, which cannot easily be seen in more realistic experiments or in real world data. The plots for all models and the entire data are shown in Fig. 5.11.

The orange line shows the point-wise morphological intelligence for the linear muscle model (MusLin) and the blue line for the non-linear muscle model (MusFib). The green line shows the point-wise morphological intelligence for the motor model (DCMot). In the figure, the lower lines show the position y of the centre of mass over time. The PD-controller of the DCMot model is parametrized to follow the trajectory of the MusFib model, which is why the blue and green position plots coincide. The original data is shown in Fig. 5.11. There are basically three phases, which need to be distinguished (indicated by the vertical lines). First, the flight phase, during which the hopper does not touch the ground (position plots are above the red line), second, the deceleration phase, which occurs after landing (position is below the red line but still declining), and finally, the acceleration phase, in which the position is below the red line but increasing.

The first observation is that morphological intelligence is very similar for all models during most of the flight phase (position above the red line) and that it is proportional to the velocity of the systems. By that we mean that morphological intelligence decreases when the velocity during flight decreases and increases when the system's speed (towards the ground) increases. During flight, the behaviour of the system is governed only by the interaction of the body (mass, velocity) and the environment (gravity) and not by the actuators. Also, all actuator control signals are constant during flight. This explains why the values coincide for the three models.

For all models, morphological intelligence drops as soon as the systems touch the ground. DCMot and MusLin reach their highest values only during the flight phase, which can be expected at least from a motor model that is not designed to exploit morphological intelligence. The graphs also reveal that the MusLin model shows slightly higher morphological intelligence around mid-stance phase, compared to the DCMot model. For the non-linear muscle model, the behaviour is different. Shortly after touching the ground, the system shows a strong decline of morphological intelligence, which is followed by a strong incline during the deceleration with the muscle. Contrary to the other two models, the non-linear muscle model MusFib shows the highest values when the muscle is contracted the most (until mid-stance). This is an interesting result, as it shows that the non-linear muscle is capable of showing more morphological intelligence while the muscle is operating, compared to the flight phase, in which the behaviour is only determined by the interaction of the body and environment.

It was mentioned in previous chapters, that the goal is to apply the proposed measure to data acquired from real-world experiments, e.g. gathered from humans. For this purpose, the second chapter introduced estimators for entropy, mutual information, and conditional mutual information on continuous state spaces (see Sect. 2.8). The experiment presented above allows us to compare how the continuous estimator compares to the discretised estimator (see Fig. 5.10). The plot on the left-hand shows the discrete estimator (see Fig. 5.11) while the plot on the right-hand side shows the results for the estimator for conditional mutual information that was presented in Sect. 2.8.1.

There three observations that need to be pointed out. First, the absolute values differ. By this we mean that the discrete estimator shows higher values than the continuous estimator. The reason is that the continuous estimator was derived for the logarithm with base e. Second, the peak for the non-linear muscle model close to time step 200, that is clearly seen for the discrete estimator is not visible for the continuous estimator. Third, the continuous estimator also allows for a point-wise estimation of MI_W. This can be seen, because the evolution of the curves for the continuous estimator (with exception of the mentioned peak) is very close to the

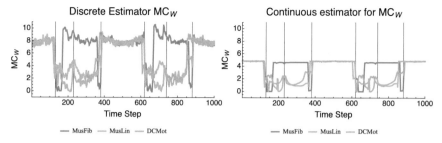

Fig. 5.10 *Comparison of discrete and continuous estimator for* MI_W *on hopping data*. This figure compares the discrete and continuous estimators for MI_W on the hopping data. The plot on the left-hand side is taken from Fig. 5.11 and shown here for better comparison. The plot on the right-hand side was generated with the continuous estimators by [20] that was discussed in Sect. 2.8.1. The comparison of both plots show that the continuous estimator captures a lot of the characteristics that a visible for the discrete estimator. Especially for the linear muscle model (MusLin) and the DC-Motor model (DCMot), the evolution of the curves is very similar. The significant difference between the two estimators is seen for the non-linear muscle model. The continuous estimator does not show the spike in morphological intelligence, that is visible for the discrete estimator

progress of the curves for the discrete estimator. This allows for the conclusion that the estimator for conditional mutual information by [20] is useful to estimate MI_W on continuous data.

5.2.4 *Discussion*

This section demonstrated the applicability of MI_W and MI_{MI} in experiments with non-trivial, biologically realistic hopping models and discussed the importance of a state-based analysis of morphological computation. The first quantification, MI_W, measures morphological intelligence as the conditional mutual information of the world and actuator states. Morphological intelligence is the additional information that the previous world state W provides about the next world state W', given that the current actuator state A is known (see Sec 3.4). The second quantification, MI_{MI}, compares the behaviour and controller complexity to determine the amount of morphological intelligence (see Sect. 3.3.1).

The numerical results of the two quantifications MI_W and MI_{MI} confirm our hypothesis that the MusFib model should show significantly higher morphological intelligence, compared to the two other models (MusLin, DCMot). These result complement previous findings showing that the minimum information required to generate hopping is reduced by the material properties of the non-linear muscle fibres compared to the DC-motor driven model [17]. More precisely, the higher morphological intelligence in the MusFib model can be attributed to the force-velocity characteristics of the muscle fibres. Previous hopping simulations showed, that these non-linear contraction dynamics reduce the influence of the controller on the actual

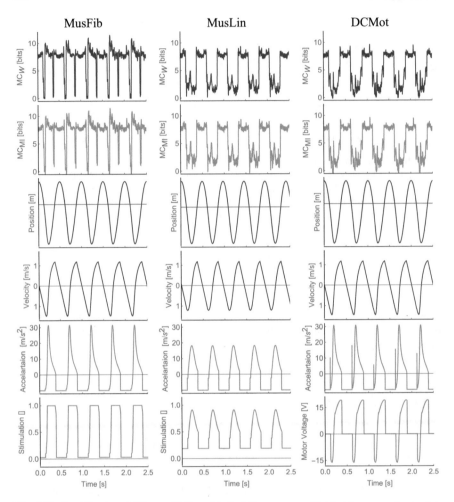

Fig. 5.11 *Visualisation of MI$_W$ and MI$_{MI}$ with the full state space.* The first two rows show point-wise MI$_W$ and point-wise MI$_W$. The lower four rows display the complete state space, i.e., the position y, velocity $\dot{y}$, and acceleration $\ddot{y}$ of the hopping systems, as well as the simulation (actuator signal). Note that the stimulation is dimensionless

hopping kinematics [16, 19]. This means, that the kinematic trajectory is more pre-determined by the material properties as compared to the linearised muscle model MusLin, or the motor model DCMot. Similar effects have been demonstrated for jumping [15], walking [21, 22]. In conclusion, this implied that studies on neural control of biological movement should consider the biomechanical characteristics of muscle contraction e.g. [23–25].

We also showed that a point-wise analysis of morphological intelligence leads to additional insights. Here we see that the non-linear muscle model is capable of showing significantly more morphological intelligence in the stance phase, compared to the flight phase, during which the behaviour is only determined by the interaction of the body and environment. This shows that morphological intelligence cannot only be assigned to the behaviour in total, but can also be analysed in a meaningful way at every point in time. Furthermore, we also show that there are methods available that allow estimating MI_W on continuous data. Future work will include the analysis of additional behaviours, such as walking and running, for which we expect, based on the findings of this section, to see more morphological intelligence of the non-linear muscle model MusFib.

To summarize the previous paragraphs, in this work we have showed that the results obtained from the two measures MI_W and MI_{MI} correspond to the intuitive understanding of morphological intelligence in muscle and DC-motor models. Furthermore, the results are in accordance with previous work on the control complexity of these models [17]. We also showed that both measures show very similar results for deterministic systems.

References

1. Zahedi K, Ay N (2013) Quantifying morphological computation. Entropy 15(5):1887–1915
2. Ghazi-Zahedi K, Haeufle DF, Montufar GF, Schmitt S, Ay N (2016) Evaluating morphological computation in muscle and dc-motor driven models of hopping movements. Front Robot AI 3(42):
3. Ghazi-Zahedi K, Deimel R, Montúfar G, Wall V, Brock O (2017a) Morphological computation: the good, the bad, and the ugly. In: 2017 IEEE/RSJ international conference on intelligent robots and systems (IROS), pp 464–469
4. Verl A, Albu-Schäffer A, Brock O, Raatz A (eds) (2015) Soft robotics: transferring theory to application. Springer
5. Hughes J, Culha U, Giardina F, Guenther F, Rosendo A, Iida F (2016) Soft manipulators and grippers: a review. Front Robot AI 3:69
6. Deimel R, Brock O (2015) A novel type of compliant and underactuated robotic hand for dexterous grasping. Int J Robot Res 35(1–3):161–185
7. Martius G, Der R, Ay N (2013) Information driven self-organization of complex robotic behaviors. PLoS ONE 8(5):e63400
8. van der Maaten L, Hinton G (2008) Visualizing high-dimensional data using t-SNE. J Mach Learn Res 9(85):2579–2605
9. Tournier M, Nesme M, Gilles B, Faure F (2015) Stable constrained dynamics. ACM Trans Graph 34(4):132:1–132:10
10. Allard J, Cotin S, Faure F, Bensoussan PJ, Poyer F, Duriez C, Delingette H, Grisoni L (2007) SOFA—an open source framework for medical simulation. Medicine meets virtual reality. Long Beach, California, Etats-Unis, pp 13–18
11. Deimel R (2017) Soft robotic hands for compliant grasping. PhD thesis, Technische Universität Berlin, Berlin
12. Clark DD, Sokoloff L (1999) Circulation and energy metabolism of the brain. In: Siegel GJ, Agranoff BW, Albers RW, Fisher SK, Uhler MD (eds) Basic neurochemistry: molecular, cellular and medical aspects, 6th edn, Lippincott-Raven, Philadelphia, chap 31
13. Sokoloff L, Mangold R, Wechsler R, Kennedy C, Kety S (1955) Effect of mental arithmetic on cerebral circulation and metabolism. J Clin Invest 34(7):1101–1108
14. Schmitt S, Haeufle DFB (2015) Mechanics and thermodynamics of biological muscle—a simple model approach. In: Verl A, Albu-Schäffer A, Brock O, Raatz A (eds) Soft Robotics, 1st edn. Springer, pp 134–144

15. van Soest AJ, Bobbert MF (1993) The contribution of muscle properties in the control of explosive movements. Biol Cybern 69(3):195–204
16. Haeufle DFB, Grimmer S, Kalveram KT, Seyfarth A (2012) Integration of intrinsic muscle properties, feed-forward and feedback signals for generating and stabilizing hopping. J Royal Soc Interface 9(72):1458–1469
17. Haeufle DFB, Günther M, Wunner G, Schmitt S (2014) Quantifying control effort of biological and technical movements: an information-entropy-based approach. Phys Rev E 89:012716
18. Ghazi-Zahedi K (2017b) Go implementations of entropy measures. http://github.com/kzahedi/goent
19. Haeufle DFB, Grimmer S, Seyfarth A (2010) The role of intrinsic muscle properties for stable hopping–stability is achieved by the force-velocity relation. Bioinspiration Biomim 5(1):016004
20. Frenzel S, Pompe B (2007) Partial mutual information for coupling analysis of multivariate time series. Phys Rev Lett 99:204101
21. Gerritsen KG, van den Bogert AJ, Hulliger M, Zernicke RF (1998) Intrinsic muscle properties facilitate locomotor control—a computer simulation study. Motor Control 2(3):206–20
22. John CT, Anderson FC, Higginson JS, Delp SL (2013) Stabilisation of walking by intrinsic muscle properties revealed in a three-dimensional muscle-driven simulation. Comput Methods Biomech Biomed Eng 16(4):451–62
23. Pinter IJ, Van Soest AJ, Bobbert MF, Smeets JBJ (2012) Conclusions on motor control depend on the type of model used to represent the periphery. Biol Cybern 106(8–9):441–451
24. Schmitt S, Günther M, Rupp T, Bayer A, Häufle D (2013) Theoretical Hill-type muscle and stability: numerical model and application. Comput Math Methods Med 2013:570878
25. Dura-Bernal S, Li K, Neymotin SA, Francis JT, Principe JC, Lytton WW (2016) Restoring behavior via inverse neurocontroller in a lesioned cortical spiking model driving a virtual arm. Front Neurosc 10(FEB):1–17

Chapter 6
Conclusions

The term morphological computation was first introduced about a decade ago to capture the observation that the exploitation of physical properties of the body and their interaction with the environment leads to a reduction of controller complexity. Two intuitive examples were given in the first chapter, grasping and running. In both cases, physical properties of the body, i.e., skin properties and elasticity of the muscle-tendon system, lead to a reduction of the amount of computation that the brain has to do. In case of grasping, the physical properties of the skin compensate less precise grasps, and in case of running, the elasticity of the muscle-tendon system compensates for the unevenness in the ground without the need for the brain to precisely monitor the ground's roughness.

Over the years, this rather vaguely defined concept has been further sharpened, where sharpened refers to a better understanding which kind of processes should be understood as physical computation conducted by the body. This has narrowed the applicability of morphological computation to a very limited set of systems. We believe the contribution of a body to intelligence is a fundamental concept in the context of embodied (artificial) intelligence that should not be understood in this narrowing and limiting way.

Hence, the goal of this work was twofold. The first goal is to develop a unifying perspective that incorporates all physical processes which lead to a reduction of the cognitive load. To develop this perspective, we first presented a large set of examples, organised in five categories, namely, morphological computation, morphological control, pre-processing in sensors, post-processing in actuators, and physical processes which do not contribute to a reduction of the cognitive load. This was then followed by a discussion of the terms *morphology* and *computation* which lead to the conclusion that the term *Morphological Computation* is not well-suited to capture the different categories. This lead to the introduction of the term *Morphological Intelligence* which we believe is better suited to describe the contribution of the body and its interaction with the environment to intelligence. The chapter closed with

© Springer Nature Switzerland AG 2019
K. Ghazi-Zahedi, *Morphological Intelligence*,
https://doi.org/10.1007/978-3-030-20621-5_6

a colloquial definition of *Morphological Intelligence* and a discussion on how this relates to the previously given examples.

The second goal of this book is to find a formal approach to quantifying morphological intelligence. The second chapter presents five different concepts of how different aspects of morphological intelligence can be quantified. Related work, the current state of the art and open questions are also discussed. Note, that most quantifications which are presented in this chapter are available as binary and source code (see Appendix A).

The following chapters in this book investigated the presented quantifications numerical (see Chap. 4) and in two applications (see Chap. 5). The numerical experiments are based on a parametrised binary model of the sensorimotor loop. This type of analysis is valuable, as it can reveal properties of the measures that cannot be seen when applied to real data and are very difficult to see from a purely analytical treatment. In the presented experiments, we varied the causal dependencies between the random variables of the sensorimotor loop from complete independence to full determinism. The numerical analysis revealed that the quantifications MI_W, MI_{CA}, and C_W are best-suited for application on real data.

The final chapter in this book presented results in which MI_W was applied to soft robotics and MI_W and MI_{MI} were applied to hopping models. The results show that although a final answer to the question of quantifying morphological intelligence has not yet been found, the currently available measures are useful in the analysis of real-world data.

This book is the first to present a quantitative approach to understanding and modelling morphological intelligence, and hence, also morphological computation. There are still open questions and problems to be solved in this field, yet, we believe that this book presents a comprehensive discussion of how the exploitation of the body and its interactions with the environment can lead to a reduction of the cognitive load. Furthermore, this book is the first to present a detailed discussion of how morphological intelligence can be quantified. The numerical simulations provide insights into the similarities and differences of the different measures. Finally, the experiments on soft robotics and muscle models prove the applicability of the current measures on real-world data. The appendix presents the software that can be used on real-world data and that was used to generate the results presented in this book.

Appendix A
gomi

gomi is a software package for the evaluation of morphological intelligence measures on data. *gomi* includes all currently available morphological intelligence measures (see Sect. 3). This includes measures with continuous estimators as well as measures that use a discrete estimator. The following sections will only provide minimal information about the measures.

gomi is written in the programming language *Go* [1]. Pre-compiled binaries for Windows, Linux, and macOS and the source code are available online [2].

A.1 Installation

For the installation of *Go*, please see [1]. Once *Go* is installed, *gomi* can easily be installed using the following commands:

```
go get github.com/kzahedi/gomi
```

The following two packages are required and might have to be installed manually:

```
go get github.com/kzahedi/goent
go get github.com/kzahedi/utils
```

© Springer Nature Switzerland AG 2019   157
K. Ghazi-Zahedi, *Morphological Intelligence*,
https://doi.org/10.1007/978-3-030-20621-5

A.2   Discrete Measures

A.2.1 MI_W

MI_W is defined in the following way (for details, see Sect. 3.4):

$$MI_W = I(W'; W|A) \tag{A.1}$$

$$= \sum_{w',w,a} p(w', w, a) \log_2 \frac{p(w'|w, a)}{p(w'|a)} \tag{A.2}$$

To calculate this measure, we need to estimate the joint distribution from which the two conditional distributions can be calculated in the following way:

$$p(w, a) = \sum_{w'} p(w', w, a) \qquad\qquad p(w', a) = \sum_{w} p(w', w, a) \tag{A.3}$$

$$p(a) = \sum_{w',w} p(w', w, a) \qquad\qquad p(w'|w, a) = \frac{p(w', w, a)}{p(w, a)} \tag{A.4}$$

$$p(w'|a) = \frac{p(w', a)}{p(a)} \tag{A.5}$$

gomi performs all required calculation automatically. As a user of *gomi*, one only needs to provide the data and some parameters, which are explained next. To calculate MI_W with the binary, use the following command line parameters:

```
gomi -mi MI_W -file musfib.csv -wi 1,2,3 -ai 9 -v \
    -bins 300 -o MI_W.csv
```

The file musfib.csv was used in [3]. It contains the data for the non-linear muscle model (see Sect. 5.2) and can be downloaded from https://github.com/kzahedi/entropy.

The command line options are

-mi MI_W	chooses MI_W as measure
-file musfib.csv	data file containing W and A
-wi 1,2,3	columns in musfib.csv that define the world state
-ai 9	columns in musfib.csv that define the actuator state
-v	will print useful information
-bins 300	determines the number of bins for W and A
-o MI_W.csv	output file, containing the results

The measures implemented in *gomi* can also be used as a library. The following code snippet gives an example:

```go
package main

import (
    "fmt"

    "github.com/kzahedi/goent/dh"
    goent "github.com/kzahedi/goent/discrete"
    mc "github.com/kzahedi/gomi/discrete"
)

func main() {
    // W and A are just examples for data.
    // These would usually be read from some data file
    // For this example to work, we provide some dummy data
    w := [][]float64{{0.0, 1.0},
        {0.1, 1.1},
        {0.2, 1.2},
        {0.3, 1.3},
        {0.4, 1.4}}

    a := [][]float64{{0.0, 1.0, 2.0},
        {0.1, 1.1, 2.1},
        {0.2, 1.2, 2.2},
        {0.3, 1.3, 2.3},
        {0.4, 1.4, 2.4}}

    // discretising data
    wDiscretised := dh.Discretise(w, []int{10, 10},
      []float64{0.0, 0.0}, []float64{1.0, 2.0})
    aDiscretised := dh.Discretise(a, []int{10, 10, 10},
      []float64{0.0, 0.0, 0.0}, []float64{1.0, 2.0, 3.0})

    // univariate variables
    wUnivariate := dh.MakeUnivariateRelabelled(wDiscretised,
      []int{10, 10})
    aUnivariate := dh.MakeUnivariateRelabelled(aDiscretised,
      []int{10, 10, 10})

    // creating w', w, a data
    w2w1a1 := make([][]int, len(w)-1, len(w)-1)

    for i := 0; i < len(w)-1; i++ {
        w2w1a1[i] = make([]int, 3, 3)
        w2w1a1[i][0] = wUnivariate[i+1]
        w2w1a1[i][1] = wUnivariate[i]
        w2w1a1[i][2] = aUnivariate[i]
    }

    // calculating p(w',w,a)
    pw2w1a1 := goent.Empirical3D(w2w1a1)

    // calculating MI_W
    result := mc.MorphologicalComputationW(pw2w1a1)

    fmt.Println(result)
}
```

A.2.2 MI_A

MI_A is defined in the following way (see Sect. 3.3):

$$MI_A = I(W'; A|W) \tag{A.6}$$

$$= \sum_{w',w,a} p(w', w, a) \log_2 \frac{p(w'|w, a)}{p(w'|w)} \tag{A.7}$$

To calculate this measure, we need to estimate the joint distributions from which the two conditional distributions can be calculated in the following way:

$$p(w, a) = \sum_{w'} p(w', w, a) \qquad\qquad p(w', w) = \sum_{a} p(w', w, a) \tag{A.8}$$

$$p(a) = \sum_{w',w} p(w', w, a) \qquad\qquad p(w'|w, a) = \frac{p(w', w, a)}{p(w, a)} \tag{A.9}$$

$$p(w'|w) = \frac{p(w', w)}{p(w)} \tag{A.10}$$

These calculations, and their estimation from data, are done automatically by *gomi*. The user only needs to provide the data and some parameters, which are explained next.

To calculate MI_A with the binary, use the following command line parameters:

```
gomi -mi MI_A -file musfib.csv -wi 1,2,3 -ai 9 -v \
    -bins 300 -o MI_A.csv
```

The file musfib.csv was used in [3]. It contains the data for the non-linear muscle model (see Sect. 5.2) and can be downloaded from https://github.com/kzahedi/entropy.

The command line options are

-mi MI_A	chooses MI_W as measure
-file musfib.csv	data file containing W and A
-wi 1,2,3	columns in musfib.csv that define the world state
-ai 9	columns in musfib.csv that define the actuator state
-v	will print useful information
-bins	300 determines the number of bins for W and A
-o MI_A.csv	output file, containing the results

The measures implemented in *gomi* can also be used as a library. The following code snippet gives an example:

```go
package main

import (
    "fmt"

    "github.com/kzahedi/goent/dh"
    goent "github.com/kzahedi/goent/discrete"
    mc "github.com/kzahedi/gomi/discrete"
)

func main() {
    // W and A are just examples for data.
    // These would usually be read from some data file
    w := [][]float64{{0.0, 1.0},
      {0.1, 1.1},
  {0.2, 1.2},
  {0.3, 1.3},
  {0.4, 1.4}}

 a := [][]float64{{0.0, 1.0, 2.0},
  {0.1, 1.1, 2.1},
  {0.2, 1.2, 2.2},
  {0.3, 1.3, 2.3},
  {0.4, 1.4, 2.4}}

 // discretising data.
    wDiscretised := dh.Discretise(w, []int{10, 10},
        []float64{0.0, 0.0}, []float64{1.0, 2.0})
    aDiscretised := dh.Discretise(a, []int{10, 10, 10},
        []float64{0.0, 0.0, 0.0}, []float64{1.0, 2.0, 3.0})

 // univariate variables
    wUnivariate := dh.MakeUnivariateRelabelled(wDiscretised,
        []int{10, 10})
    aUnivariate := dh.MakeUnivariateRelabelled(aDiscretised,
        []int{10, 10, 10})

 // creating w', w, a data
 w2a1w1 := make([][]int, len(w)-1, len(w)-1)

 for i := 0; i < len(w)-1; i++ {
  w2a1w1[i] = make([]int, 3, 3)
  w2a1w1[i][0] = wUnivariate[i+1]
  w2a1w1[i][1] = aUnivariate[i]
  w2a1w1[i][2] = wUnivariate[i]
 }

 // calculating p(w',a,w)
 pw2a1w1 := goent.Empirical3D(w2a1w1)

 // calculating MI_A
 result := mc.MorphologicalComputationA(pw2a1w1)

 fmt.Println(result)
}
```

A.2.3 MI'_A

$\mathrm{MI_A}$ is defined in the following way (see Sect. 3.3):

$$\mathrm{MI'_A} = 1 - \frac{1}{\log_2 |W|} I(W'; A|W) \tag{A.11}$$

$$= 1 - \frac{1}{\log_2 |W|} \sum_{w',w,a} p(w', w, a) \log_2 \frac{p(w'|w, a)}{p(w'|w)} \tag{A.12}$$

To calculate this measure, we need to estimate the joint distributions from which the two conditional distributions can be calculated in the following way:

$$p(w, a) = \sum_{w'} p(w', w, a) \qquad p(w', w) = \sum_{a} p(w', w, a) \tag{A.13}$$

$$p(a) = \sum_{w',w} p(w', w, a) \qquad p(w'|w, a) = \frac{p(w', w, a)}{p(w, a)} \tag{A.14}$$

$$p(w'|w) = \frac{p(w', w)}{p(w)} \tag{A.15}$$

These calculations, and their estimation from data, are conducted automatically by *gomi*. The user only needs to provide the data and some parameters, which are explained next.

To calculate $\mathrm{MI'_A}$ with the binary, use the following command line parameters:

```
gomi -mi MI_A_Prime -file musfib.csv -wi 1,2,3 -ai 9 -v \
   -bins 300 -o MI_A_Prime.csv
```

The file musfib.csv was used in [3]. It contains the data for the non-linear muscle model (see Sect. 5.2) and can be downloaded from https://github.com/kzahedi/entropy.

The command line options are

-mi MI_A_Prime	chooses $\mathrm{MI_W}$ as measure
-file musfib.csv	data file containing W and A
-wi 1,2,3	columns in musfib.csv that define the world state
-ai 9	columns in musfib.csv that define the actuator state
-v	will print useful information
-bins 300	determines the number of bins for W and A
-o MI_A_Prime.csv	output file, containing the results

The measures implemented in *gomi* can also be used as a library. The following code snippet gives an example:

```go
import (
    "fmt"
    "math"

    "github.com/kzahedi/goent/dh"
    goent "github.com/kzahedi/goent/discrete"
    mc "github.com/kzahedi/gomi/discrete"
)

func main() {
    // W and A are just examples for data.
    // These would usually be read from some data file
    w := [][]float64{{0.0, 1.0},
        {0.1, 1.1},
        {0.2, 1.2},
        {0.3, 1.3},
        {0.4, 1.4}}

    a := [][]float64{{0.0, 1.0, 2.0},
        {0.1, 1.1, 2.1},
        {0.2, 1.2, 2.2},
        {0.3, 1.3, 2.3},
    {0.4, 1.4, 2.4}}

    // discretising data.
    wDiscretised := dh.Discretise(w, []int{10, 10},
        []float64{0.0, 0.0}, []float64{1.0, 2.0})
    aDiscretised := dh.Discretise(a, []int{10, 10, 10},
        []float64{0.0, 0.0, 0.0}, []float64{1.0, 2.0, 3.0})

    // univariate variables
    wUnivariate := dh.MakeUnivariateRelabelled(wDiscretised,
        []int{10, 10})
    aUnivariate := dh.MakeUnivariateRelabelled(aDiscretised,
        []int{10, 10, 10})

    // creating w', w, a data
    w2a1w1 := make([][]int, len(w)-1, len(w)-1)

    for i := 0; i < len(w)-1; i++ {
        w2a1w1[i] = make([]int, 3, 3)
        w2a1w1[i][0] = wUnivariate[i+1]
        w2a1w1[i][1] = aUnivariate[i]
        w2a1w1[i][2] = wUnivariate[i]
    }

    // calculating p(w',a,w)
    pw2a1w1 := goent.Empirical3D(w2a1w1)
    abins := 10.0 * 10.0 * 10.0 // see aDiscretised above

    // calculating MI_A_Prime
    result := 1.0 - 1.0/math.Log2(abins)
        - mc.MorphologicalComputationA(pw2a1w1)

    fmt.Println(result)
}
```

A.2.4 MI_MI

$\mathrm{MI_{MI}}$ is defined in the following way (see Sect. 3.3.1):

$$\mathrm{MI_{MI}} = I(W'; W) - I(A; S) \tag{A.16}$$

$$= \sum_{w', w} p(w', w) \log_2 \frac{p(w', w)}{p(w')p(w)} - \sum_{a, s} p(a, s) \log_2 \frac{p(a, s)}{p(a)p(s)} \tag{A.17}$$

To calculate $\mathrm{MI_{MI}}$ with the binary, use the following command line parameters:

```
gomi -mi MI_MI -file musfib.csv -wi 1,2,3 -ai 9 -si 4 -v \
    -bins 300 -o MI_MI.csv
```

The file musfib.csv was used in [3]. It contains the data for the non-linear muscle model (see Sect. 5.2) and can be downloaded from https://github.com/kzahedi/entropy.

The command line options are

-mi MI_MI	chooses $\mathrm{MI_W}$ as measure
-file musfib.csv	data file containing W and A
-wi 1,2,3	columns in musfib.csv that define the world state
-ai 9	columns in musfib.csv that define the actuator state
-si 4	columns in musfib.csv that define the sensor state
-v	will print useful information
-bins 300	determines the number of bins for W and A
-o MI_MI.csv	output file, containing the results

The measures implemented in *gomi* can also be used as a library. The following code snippet gives an example:

```go
import (
    "fmt"

    "github.com/kzahedi/goent/dh"
    goent "github.com/kzahedi/goent/discrete"
    mc "github.com/kzahedi/gomi/discrete"
)

func main() {
    // W, A, and S are just examples for data.
    // These would usually be read from some data file
    w := [][]float64{{0.0, 1.0},
        {0.1, 1.1},
        {0.2, 1.2},
        {0.3, 1.3},
        {0.4, 1.4}}

    a := [][]float64{{0.0, 1.0, 2.0},
        {0.1, 1.1, 2.1},
        {0.2, 1.2, 2.2},
```

```
        {0.3,  1.3,  2.3},
        {0.4,  1.4,  2.4}}

s := [][]float64{{0.0},
      {0.1},
      {0.2},
      {0.3},
      {0.4}}

// discretising data.
wDiscretised := dh.Discretise(w, []int{10, 10},
    []float64{0.0, 0.0}, []float64{1.0, 2.0})
aDiscretised := dh.Discretise(a, []int{10, 10, 10},
    []float64{0.0, 0.0, 0.0}, []float64{1.0, 2.0, 3.0})
sDiscretised := dh.Discretise(s, []int{10},
    []float64{0.0}, []float64{1.0})

// univariate variables
wUnivariate := dh.MakeUnivariateRelabelled(wDiscretised,
    []int{10, 10})
aUnivariate := dh.MakeUnivariateRelabelled(aDiscretised,
    []int{10, 10, 10})
sUnivariate := dh.MakeUnivariateRelabelled(sDiscretised,
    []int{10})

// creating w', w and a, s data
w2w1 := make([][]int, len(w)-1, len(w)-1)
a1s1 := make([][]int, len(w)-1, len(w)-1)

for i := 0; i < len(w)-1; i++ {
    w2w1[i] = make([]int, 2, 2)
    w2w1[i][0] = wUnivariate[i+1]
    w2w1[i][1] = wUnivariate[i]
    a1s1[i] = make([]int, 2, 2)
    a1s1[i][0] = aUnivariate[i]
    a1s1[i][1] = sUnivariate[i]
}

// calculating p(w',w) ,and p(a,s)
pw2w1 := goent.Empirical2D(w2w1)
pa1s1 := goent.Empirical2D(w2w1)

// calculating MI_MI
result := mc.MorphologicalComputationMI(pw2w1, pa1s1)

fmt.Println(result)
}
```

A.2.5 MI_{CA}

MI_{CA} was defined in the following way (see Sect. 3.3.2):

$$MI_{CA} = CIF(W \to W') - CIF(A \to W') \tag{A.18}$$

$$= I(W'; W) - I(W'; A) \tag{A.19}$$

$$= \sum_{w', w} p(w', w) \log_2 \frac{p(w', w)}{p(w')p(w)} - \sum_{w', a} p(w', a) \log_2 \frac{p(w', a)}{p(w')p(a)} \tag{A.20}$$

To calculate MI_{CA} with the binary, use the following command line parameters:

```
gomi -mi MI_CA -file musfib.csv -wi 1,2,3 -ai 9 -v \
    -bins 300 -o MI_CA.csv
```

The file musfib.csv was used in [3]. It contains the data for the non-linear muscle model (see Sect. 5.2) and can be downloaded from https://github.com/kzahedi/entropy.

The command line options are

-mi MI_CA	chooses MI_W as measure
-file musfib.csv	data file containing W and A
-wi 1,2,3	columns in musfib.csv that define the world state
-ai 9	columns in musfib.csv that define the actuator state
-v	will print useful information
-bins 300	determines the number of bins for W and A
-o MI_CA.csv	output file, containing the results

The measures implemented in *gomi* can also be used as a library. The following code snippet gives an example:

```
package main

import (
    "fmt"

    "github.com/kzahedi/goent/dh"
    goent "github.com/kzahedi/goent/discrete"
    mc "github.com/kzahedi/gomi/discrete"
)

func main() {
    // W and A are just examples for data.
    // These would usually be read from some data file
    w := [][]float64{{0.0, 1.0},
        {0.1, 1.1},
        {0.2, 1.2},
        {0.3, 1.3},
        {0.4, 1.4}}
```

```
a := [][]float64{{0.0,  1.0,  2.0},
     {0.1,  1.1,  2.1},
     {0.2,  1.2,  2.2},
     {0.3,  1.3,  2.3},
     {0.4,  1.4,  2.4}}

// discretising data.
wDiscretised := dh.Discretise(w, []int{10, 10},
     []float64{0.0,  0.0}, []float64{1.0,  2.0})
aDiscretised := dh.Discretise(a, []int{10, 10, 10},
     []float64{0.0, 0.0, 0.0}, []float64{1.0, 2.0, 3.0})

// univariate variables
wUnivariate := dh.MakeUnivariateRelabelled(wDiscretised,
     []int{10, 10})
aUnivariate := dh.MakeUnivariateRelabelled(aDiscretised,
     []int{10, 10, 10})

// creating w', w, and w', a data
w2w1 := make([][]int, len(w)-1, len(w)-1)
w2a1 := make([][]int, len(w)-1, len(w)-1)

for i := 0; i < len(w)-1; i++ {
    w2w1[i] = make([]int, 2, 2)
    w2w1[i][0] = wUnivariate[i+1]
    w2w1[i][1] = wUnivariate[i]
    w2a1[i] = make([]int, 2, 2)
    w2a1[i][0] = wUnivariate[1+1]
    w2a1[i][1] = aUnivariate[i]
}

// calculating p(w',w,a)
pw2w1 := goent.Empirical2D(w2w1)
pw2a1 := goent.Empirical2D(w2a1)

// calculating MI_CA
result := mc.MorphologicalComputationCA(pw2w1, pw2a1)

fmt.Println(result)
}
```

A.2.6 C_A

C_A is defined in the following way (see Sect. 3.3):

$$C_A := 1 + \frac{1}{\log |\mathcal{S}|}(CIF(S \to S') - CIF(A \to S')) \tag{A.21}$$

$$= 1 - \frac{1}{\log |\mathcal{S}|} \sum_{s,a} p(s,a) \sum_{s'} p(s'|\text{do}(a)) \log \frac{p(s'|\text{do}(a))}{p(s'|\text{do}(s))} \tag{A.22}$$

The starting point for the calculation C_A is the joint distribution $p(s', s, a)$, from which the other distributions can be calculated in the following way:

$$p(s) = \sum_{s',a} p(s', s, a) \qquad\qquad p(s, a) = \sum_{s'} p(s', s, a) \qquad\qquad (A.23)$$

$$p(s'|s, a) = \frac{p(s', s, a)}{p(s, a)} \qquad\qquad p(a|s) = \frac{p(s, a)}{p(s)} \qquad\qquad (A.24)$$

$$p(s'|\mathrm{do}(a)) = \sum_{s} p(s'|s, a)p(s) \quad p(s'|\mathrm{do}(s)) = \sum_{a} p(a|s)p(s'|\mathrm{do}(a)) \quad (A.25)$$

To calculate CA with the binary, use the following command line parameters:

```
gomi -mi CA -file musfib.csv -si 4 -ai 9 -v \
     -bins 300 -o CA.csv
```

The file musfib.csv was used in [3]. It contains the data for the non-linear muscle model (see Sect. 5.2) and can be downloaded from https://github.com/kzahedi/entropy.

The command line options are

-mi CA	chooses MI_W as measure
-file musfib.csv	data file containing W and A
-si 4	columns in musfib.csv that define the sensor state
-ai 9	columns in musfib.csv that define the actuator state
-v	will print useful information
-bins 300	determines the number of bins for S and A
-o CA.csv	output file, containing the results

The measures implemented in *gomi* can also be used as a library. The following code snippet gives an example:

```
package main

import (
    "fmt"

    "github.com/kzahedi/goent/dh"
    goent "github.com/kzahedi/goent/discrete"
    mc "github.com/kzahedi/gomi/discrete"
)

func main() {
    // S and A are just examples for data.
    // These would usually be read from some data file
    s := [][]float64{{0.0, 1.0},
        {0.1, 1.1},
        {0.2, 1.2},
        {0.3, 1.3},
        {0.4, 1.4}}
```

```
a := [][]float64{{0.0,  1.0,  2.0},
    {0.1,  1.1,  2.1},
    {0.2,  1.2,  2.2},
    {0.3,  1.3,  2.3},
    {0.4,  1.4,  2.4}}

// discretising data
sDiscretised := dh.Discretise(s, []int{10, 10},
    []float64{0.0, 0.0}, []float64{1.0, 2.0})
aDiscretised := dh.Discretise(a, []int{10, 10, 10},
    []float64{0.0, 0.0, 0.0}, []float64{1.0, 2.0, 3.0})

// univariate variables
sUnivariate := dh.MakeUnivariateRelabelled(sDiscretised,
    []int{10, 10})
aUnivariate := dh.MakeUnivariateRelabelled(aDiscretised,
    []int{10, 10, 10})

// creating s', s, a data
s2s1a1 := make([][]int, len(s)-1, len(s)-1)

for i := 0; i < len(a)-1; i++ {
    s2s1a1[i] = make([]int, 3, 3)
    s2s1a1[i][0] = sUnivariate[i+1]
    s2s1a1[i][1] = sUnivariate[i]
    s2s1a1[i][2] = aUnivariate[i]
}

// calculating p(s',s,a)
ps2s1a1 := goent.Empirical3D(s2s1a1)
sbins := 10 * 10 // see sDiscretised above

// calculating CA
result := mc.MorphologicalComputationIntrinsicCA(ps2s1a1,
    sbins)

fmt.Println(result)
}
```

A.2.7 MI_{SY}

To calculate MI_{SY} with the binary, use the following command line parameters:

```
gomi -mi MI_SY -file musfib.csv -wi 1,2,3 -ai 9 -v \
    -bins 300 -o MI_SY.csv
```

The file musfib.csv was used in [3]. It contains the data for the non-linear muscle model (see Sect. 5.2) and can be downloaded from https://github.com/kzahedi/entropy.

The command line options are

-mi MI_SY A	chooses MI_W as measure
-file musfib.csv	data file containing W and A
-wi 1,2,3	columns in musfib.csv that define the world state
-ai 9	columns in musfib.csv that define the actuator state
-v	will print useful information
-bins 300	determines the number of bins for S and A
-o MI_SY.csv	output file, containing the results

The measures implemented in *gomi* can also be used as a library. The following code snippet is an example of how it can be used:

```
package main

import (
    "fmt"

    "github.com/kzahedi/goent/dh"
    goent "github.com/kzahedi/goent/discrete"
    mc "github.com/kzahedi/gomi/discrete"
)

func main() {
    // W and A are just examples for data.
    // These would usually be read from some data file
    w := [][]float64{{0.0, 1.0},
        {0.1, 1.1},
        {0.2, 1.2},
        {0.3, 1.3},
        {0.4, 1.4}}

    a := [][]float64{{0.0, 1.0, 2.0},
        {0.1, 1.1, 2.1},
        {0.2, 1.2, 2.2},
        {0.3, 1.3, 2.3},
        {0.4, 1.4, 2.4}}

    // discretising data
    wDiscretised := dh.Discretise(w, []int{10, 10},
        []float64{0.0, 0.0}, []float64{1.0, 2.0})
    aDiscretised := dh.Discretise(a, []int{10, 10, 10},
        []float64{0.0, 0.0, 0.0}, []float64{1.0, 2.0, 3.0})

    // univariate variables
    wUnivariate := dh.MakeUnivariateRelabelled(wDiscretised,
        []int{10, 10})
    aUnivariate := dh.MakeUnivariateRelabelled(aDiscretised,
        []int{10, 10, 10})

    // creating w', w, a data
    w2w1a1 := make([][]int, len(w)-1, len(w)-1)

    for i := 0; i < len(w)-1; i++ {
```

```
        w2w1a1[i] = make([]int, 3, 3)
        w2w1a1[i][0] = wUnivariate[i+1]
        w2w1a1[i][1] = wUnivariate[i]
        w2w1a1[i][2] = aUnivariate[i]
    }

    // calculating p(w',w,a)
    pw2w1a1 := goent.Empirical3D(w2w1a1)

    // calculating MI_W
    iterations := 1000
    verbose := true
    result := mc.MorphologicalComputationSY(pw2w1a1,
        iterations, verbose)

    fmt.Println(result)
}
```

A.3   Continuous Measures

A.3.1 MI_W

This implementation of MI_W uses the Frenzel-Pompe estimator on continuous data
(see Sect. 2.8.2). To calculate MI_W with the binary, use the following command line
parameters:

```
gomi -mi MI_W -file musfib.csv -wi 1,2,3 -ai 9 -v -c \
    -o MI_W.csv
```

The file musfib.csv was used in [3]. It contains the data for the non-linear muscle model (see Sect. 5.2) and can be downloaded from https://github.com/kzahedi/
entropy.

The command line options are

-mi MI_W A	chooses MI_W as measure
-file musfib.csv	data file containing W and A
-wi 1,2,3	columns in musfib.csv that define the world state
-ai 9	columns in musfib.csv that define the actuator state
-v	will print useful information
-c	will use the Frenzel-Pompe estimator for continuous data
-bins 300	determines the number of bins for S and A
-o MI_W.csv	output file, containing the results

The measures implemented in *gomi* can also be used as a library. The following
code snippet gives is an example:

```go
package main

import (
    "fmt"
    "math/rand"

    goent "github.com/kzahedi/goent/continuous"
    mc "github.com/kzahedi/gomi/continuous"
)

func main() {
    // W and A are just examples for data.
    // These would usually be read from some data file
    wDim := 2
    w := make([][]float64, 1000, 1000)
    for i := 0; i < 1000; i++ {
        w[i] = make([]float64, wDim, wDim)
        w[i][0] = rand.Float64()
        w[i][1] = rand.Float64()
    }

    aDim := 3
    a := make([][]float64, 1000, 1000)
    for i := 0; i < 1000; i++ {
        a[i] = make([]float64, aDim, aDim)
        a[i][0] = rand.Float64()
        a[i][1] = rand.Float64()
        a[i][2] = rand.Float64()
    }

    // creating w', w, a data
    w2w1a1 := make([][]float64, 999, 999)

    for i := 0; i < len(w)-1; i++ {
        w2w1a1[i] = make([]float64, 2*wDim+aDim, 2*wDim+aDim)

        // w'
        w2w1a1[i][0] = w[i+1][0]
        w2w1a1[i][1] = w[i+1][1]

        // w
        w2w1a1[i][2] = w[i][0]
        w2w1a1[i][3] = w[i][1]

        // a
        w2w1a1[i][4] = a[i][0]
        w2w1a1[i][5] = a[i][1]
        w2w1a1[i][6] = a[i][2]
    }

    w2w1a1 = goent.Normalise(w2w1a1, false)

    // calculating MI_W
    w2indices := []int{0, 1}
    w1indices := []int{2, 3}
    a1indices := []int{4, 5, 6}
    k := 30
```

```
      verbose := true
      result := mc.MorphologicalComputationW(w2w1a1,
          w2indices, w1indices, a1indices, k, verbose)

    fmt.Println(result)
}
```

A.3.2   MI$_A$

This implementation of MI$_A$ uses the Frenzel-Pompe estimator on continuous data
(see Sect. 2.8.2).

To calculate MI$_A$ with the binary, use the following command line parameters:

```
gomi -mi MI_A -file musfib.csv -wi 1,2,3 -ai 9 -v -c \
    -o MI_A.csv
```

The file musfib.csv was used in [3]. It contains the data for the non-linear muscle model (see Sect. 5.2) and can be downloaded from https://github.com/kzahedi/
entropy.

The command line options are

-mi MI_W A	chooses MI$_W$ as measure
-file musfib.csv	data file containing W and A
-wi 1,2,3	columns in musfib.csv that define the world state
-ai 9	columns in musfib.csv that define the actuator state
-v	will print useful information
-c	will use the Frenzel-Pompe estimator for continuous data
-bins 300	determines the number of bins for S and A
-o MI_W.csv	output file, containing the results

The measures implemented in *gomi* can also be used as a library. The following
code snippet gives an example:

```
package main

import (
    "fmt"
    "math/rand"

    goent "github.com/kzahedi/goent/continuous"
    mc "github.com/kzahedi/gomi/continuous"
)

func main() {
    // W and A are just examples for data.
    // These would usually be read from some data file
    wDim := 2
    w := make([][]float64, 1000, 1000)
```

```
for i := 0; i < 1000; i++ {
    w[i] = make([]float64, wDim, wDim)
    w[i][0] = rand.Float64()
    w[i][1] = rand.Float64()
}

aDim := 3
a := make([][]float64, 1000, 1000)
for i := 0; i < 1000; i++ {
    a[i] = make([]float64, aDim, aDim)
    a[i][0] = rand.Float64()
    a[i][1] = rand.Float64()
    a[i][2] = rand.Float64()
}

// creating w', w, a data
w2w1a1 := make([][]float64, 999, 999)

for i := 0; i < len(w)-1; i++ {
    w2w1a1[i] = make([]float64, 2*wDim+aDim, 2*wDim+aDim)

    // w'
    w2w1a1[i][0] = w[i+1][0]
    w2w1a1[i][1] = w[i+1][1]

    // w
    w2w1a1[i][2] = w[i][0]
    w2w1a1[i][3] = w[i][1]

    // a
    w2w1a1[i][4] = a[i][0]
    w2w1a1[i][5] = a[i][1]
    w2w1a1[i][6] = a[i][2]
}

w2w1a1 = goent.Normalise(w2w1a1, false)

// calculating MI_W
w2indices := []int{0, 1}
w1indices := []int{2, 3}
a1indices := []int{4, 5, 6}
k := 30
verbose := true
result := mc.MorphologicalComputationA(w2w1a1,
    w2indices, w1indices, a1indices, k, verbose)

fmt.Println(result)
}
```

A.3.3   MI$_{MI}$

MI$_{MI}$ is defined as the difference between the two mutual informations $I(W'; W)$ and $I(A; S)$ (see Sect. 3.3.1). Both mutual informations are estimated either with

KSG1 or KSG2 (see Sect. 2.8.1). To calculate MI_{MI} with the binary, use the following command line parameters:

```
gomi -mi MI_MI -file musfib.csv -wi 1,2,3 -ai 9 -si 4 -v \
    -c -cm 2 -o MI_MI.csv
```

The file musfib.csv was used in [3]. It contains the data for the non-linear muscle model (see Sect. 5.2) and can be downloaded from https://github.com/kzahedi/entropy.

The command line options are

-mi MI_MI A	chooses MI_W as measure
-file musfib.csv	data file containing W and A
-wi 1,2,3	columns in musfib.csv that define the world state
-ai 9	columns in musfib.csv that define the actuator state
-si 4	columns in musfib.csv that define the sensor state
-v	will print useful information
-c	will use the KSG estimator for continuous data
-cm 2	continuous mode: 1 = KSG1, 2 = KSG2
-bins 300	determines the number of bins for S and A
-o MI_MI.csv	output file, containing the results

The full lists of options is given below.

The measures implemented in *gomi* can also be used as a library. The following code snippet is an example:

```
package main

import (
    "fmt"
    "math/rand"

    goent "github.com/kzahedi/goent/continuous"
    mc "github.com/kzahedi/gomi/continuous"
)

func main() {
    // W and A are just examples for data.
    // These would usually be read from some data file
    wDim := 2
    w := make([][]float64, 1000, 1000)
    for i := 0; i < 1000; i++ {
        w[i] = make([]float64, wDim, wDim)
        w[i][0] = rand.Float64()
        w[i][1] = rand.Float64()
    }

    aDim := 3
    a := make([][]float64, 1000, 1000)
    for i := 0; i < 1000; i++ {
        a[i] = make([]float64, aDim, aDim)
        a[i][0] = rand.Float64()
        a[i][1] = rand.Float64()
```

```
        a[i][2] = rand.Float64()
}

sDim := 5
s := make([][]float64, 1000, 1000)
for i := 0; i < 1000; i++ {
    s[i] = make([]float64, sDim, sDim)
    s[i][0] = rand.Float64()
    s[i][1] = rand.Float64()
    s[i][2] = rand.Float64()
    s[i][3] = rand.Float64()
    s[i][4] = rand.Float64()
}

// creating w', w, a data
w2w1s1a1 := make([][]float64, 999, 999)

for i := 0; i < len(w)-1; i++ {
    w2w1s1a1[i] = make([]float64,
        2*wDim+sDim+aDim, 2*wDim+sDim+aDim)

    // w'
    w2w1s1a1[i][0] = w[i+1][0]
    w2w1s1a1[i][1] = w[i+1][1]

    // w
    w2w1s1a1[i][2] = w[i][0]
    w2w1s1a1[i][3] = w[i][1]

    // s
    w2w1s1a1[i][4] = s[i][0]
    w2w1s1a1[i][5] = s[i][1]
    w2w1s1a1[i][6] = s[i][2]
    w2w1s1a1[i][7] = s[i][3]
    w2w1s1a1[i][8] = s[i][4]

    // a
    w2w1s1a1[i][9] = a[i][0]
    w2w1s1a1[i][10] = a[i][1]
    w2w1s1a1[i][11] = a[i][2]
}

w2w1s1a1 = goent.Normalise(w2w1s1a1, false)

// calculating MI_MI
w2indices := []int{0, 1}
w1indices := []int{2, 3}
s1indices := []int{4, 5, 6, 7, 8}
a1indices := []int{9, 10, 11}
k := 30
verbose := true

result1 := mc.MorphologicalComputationMI1(w2w1s1a1,
    w2indices, w1indices, s1indices, a1indices,
    k, verbose)
fmt.Println(result1)

result2 := mc.MorphologicalComputationMI2(w2w1s1a1,
    w2indices, w1indices, s1indices, a1indices,
```

```
            k, verbose)
      fmt.Println(result2)
}
```

A.3.4 MI_{CA}

MI_{CA} was defined as the difference between $I(W'; W)$ and $I(W'; A)$ (see Sect. 3.3.2). Both, $I(W'; W)$ and $I(W'; A)$, can be estimated with the KSG1 and KSG2 estimator for mutual information on continuous data. To calculate MI_{CA} with the binary, use the following command line parameters:

```
gomi -mi MI_CA -file musfib.csv -wi 1,2,3 -ai 9 -v -c \
   -cm 2 -o MI_CA.csv
```

The file musfib.csv was used in [3]. It contains the data for the non-linear muscle model (see Sect. 5.2) and can be downloaded from https://github.com/kzahedi/entropy.

The command line options are

-mi MI_CA	chooses MI_W as measure
-file musfib.csv	data file containing W and A
-wi 1,2,3	columns in musfib.csv that define the world state
-ai 9	columns in musfib.csv that define the actuator state
-v	will print useful information
-c	will use the KSG estimator for continuous data
-cm 2	continuous mode: 1 = KSG1, 2 = KSG2
-bins 300	determines the number of bins for W and A
-o MI_CA.csv	output file, containing the results

The measures implemented in *gomi* can also be used as a library. The following code snippet gives an example:

```
package main

import (
    "fmt"
    "math/rand"

    goent "github.com/kzahedi/goent/continuous"
    mc "github.com/kzahedi/gomi/continuous"
)

func main() {
    // W and A are just examples for data.
    // These would usually be read from some data file
    wDim := 2
    w := make([][]float64, 1000, 1000)
    for i := 0; i < 1000; i++ {
```

```go
        w[i] = make([]float64, wDim, wDim)
        w[i][0] = rand.Float64()
        w[i][1] = rand.Float64()
    }

    aDim := 3
    a := make([][]float64, 1000, 1000)
    for i := 0; i < 1000; i++ {
        a[i] = make([]float64, aDim, aDim)
        a[i][0] = rand.Float64()
        a[i][1] = rand.Float64()
        a[i][2] = rand.Float64()
    }

    // creating w', w, a data
    w2w1a1 := make([][]float64, 999, 999)

    for i := 0; i < len(w)-1; i++ {
        w2w1a1[i] = make([]float64, 2*wDim+aDim, 2*wDim+aDim)

        // w'
        w2w1a1[i][0] = w[i+1][0]
        w2w1a1[i][1] = w[i+1][1]

        // w
        w2w1a1[i][2] = w[i][0]
        w2w1a1[i][3] = w[i][1]

        // a
        w2w1a1[i][4] = a[i][0]
        w2w1a1[i][5] = a[i][1]
        w2w1a1[i][6] = a[i][2]
    }

    w2w1a1 = goent.Normalise(w2w1a1, false)

    // calculating MI_CA
    w2indices := []int{0, 1}
    w1indices := []int{2, 3}
    a1indices := []int{4, 5, 6}
    k := 30
    verbose := true

    result1 := mc.MorphologicalComputationCA1(w2w1a1,
        w2indices, w1indices, a1indices, k, verbose)
    fmt.Println(result1)

    result2 := mc.MorphologicalComputationCA2(w2w1a1,
        w2indices, w1indices, a1indices, k, verbose)
    fmt.Println(result2)
}
```

A.4   Full List of Command Line Parameters

-help   Will show command line parameters with explanations.
-v   Verbose. If not provided, gomi will be executed silently.
-mi   String identifier for the morphological intelligence quantification. One example is MI_W. For a full list, please use the help command line option.
-c   The discrete (frequency based) estimators are used by default. If this options is provided, the estimators on continuous data (KSG Estimator and Frenzel-Pompe) are used instead.
-cm   Continuous mode. There are two different KSG Estimators for mutual information. For those continuous morphological intelligence quantification which operate with mutual information, the KSG Estimator can be chosen with this command line options. Possible values are 1 and 2.
-s   State-dependent results. By default, the averaged results are provided. If this option is given, the results are calculated for each state, i.e., the result is a vector with a value for each row in the original dataset.
-bins   This option is used only for the discrete estimators. It determines the number of bins for each column of the W, S, and A datasets.
-i   Iteration. This command line option is only used for MI_SY, MI_Wp, and MI_SY_NID. It determines the number of iterations for the iterative scaling algorithm.
-o   Output file. The default value is "out.csv". gomi writes all results to the output file, including a header section that includes the full parametrisation of the calculation.
-wbins   Only used for discrete measures. In case the world state is given by more than one column in the dataset, this option allows to set the binning for each column.
-abins   Only used for discrete measures. In case the action state is given by more than one column in the dataset, this option allows to set the binning for each column.
-sbins   Only used for discrete measures. In case the sensor state is given by more than one column in the dataset, this option allows to set the binning for each column.
-file   This option should be used if W, S, A are provided in a single file. The columns for W, S, and A are the specified with the -wi, -si, and -ai options (see below).
-wi   The values here specify the column of the file provided with -file that relate to the world state.
-ai   The values here specify the column of the file provided with -file that relate to the action state.
-si   The values here specify the column of the file provided with -file that relate to the sensor state.
-wfile   World state file. All columns of this file will be considered in the calculations.

-afile   Action state file. All columns of this file will be considered in the calculations.

-sfile   Sensor state file. All columns of this file will be considered in the calculations.

-dfile   Domain file. This option will only be used for discrete measures. During discretisation, each column is normalised. For comparability between different files, a domain file can be specified, which provides the minimum and maximum values for the world state, action state, and sensor state data.

-k   This parameter is only used for continuous estimators. It is the parameter used for the k-nearest neighbour estimation of entropies.

-log   If this parameter is provided, gomi will write a JSON file that included all information with respect to calculation, including the full raw and pre-processed data.

-sparse   This parameter enables calculation of the discrete measures on the sparse matrix implementation.

References

1. Griesemer R, Pike R, Thompson K (2019) The go programming lanauge. https://golang.org
2. Ghazi-Zahedi K (2019) gomi GitHub Repository. https://github.com/kzahedi/gomi
3. Ghazi-Zahedi K, Haeufle DF, Montufar GF, Schmitt S, Ay N (2016) Evaluating morphological computation in muscle and dc-motor driven models of hopping movements. Front Robot AI 3(42)

Printed in the United States
By Bookmasters